Pia Regina Kieninger

Urbanes Engagement in der traditionellen Landbewirtschaftung

Pia Regina Kieninger

Urbanes Engagement in der traditionellen Landbewirtschaftung

Kulturlandschaftserhaltung in Japan – Eine Fallstudie

Südwestdeutscher Verlag für Hochschulschriften

Impressum / Imprint

Bibliografische Information der Deutschen Nationalbibliothek: Die Deutsche Nationalbibliothek verzeichnet diese Publikation in der Deutschen Nationalbibliografie; detaillierte bibliografische Daten sind im Internet über http://dnb.d-nb.de abrufbar.

Alle in diesem Buch genannten Marken und Produktnamen unterliegen warenzeichen-, marken- oder patentrechtlichem Schutz bzw. sind Warenzeichen oder eingetragene Warenzeichen der jeweiligen Inhaber. Die Wiedergabe von Marken, Produktnamen, Gebrauchsnamen, Handelsnamen, Warenbezeichnungen u.s.w. in diesem Werk berechtigt auch ohne besondere Kennzeichnung nicht zu der Annahme, dass solche Namen im Sinne der Warenzeichen- und Markenschutzgesetzgebung als frei zu betrachten wären und daher von jedermann benutzt werden dürften.

Bibliographic information published by the Deutsche Nationalbibliothek: The Deutsche Nationalbibliothek lists this publication in the Deutsche Nationalbibliografie; detailed bibliographic data are available in the Internet at http://dnb.d-nb.de.

Any brand names and product names mentioned in this book are subject to trademark, brand or patent protection and are trademarks or registered trademarks of their respective holders. The use of brand names, product names, common names, trade names, product descriptions etc. even without a particular marking in this works is in no way to be construed to mean that such names may be regarded as unrestricted in respect of trademark and brand protection legislation and could thus be used by anyone.

Coverbild / Cover image: www.ingimage.com

Verlag / Publisher:
Südwestdeutscher Verlag für Hochschulschriften
ist ein Imprint der / is a trademark of
AV Akademikerverlag GmbH & Co. KG
Heinrich-Böcking-Str. 6-8, 66121 Saarbrücken, Deutschland / Germany
Email: info@svh-verlag.de

Herstellung: siehe letzte Seite /
Printed at: see last page
ISBN: 978-3-8381-3549-6

Zugl. / Approved by: Wien, BOKU, Diss., 2012

Copyright © 2013 AV Akademikerverlag GmbH & Co. KG
Alle Rechte vorbehalten. / All rights reserved. Saarbrücken 2013

Rahmenschrift Dissertation Pia R. Kieninger

Abb. 1 ReisfeldpächterInnen, sogenannte *tanada ōnā* (*ōnā*, vom Englischen 'owner'), in Ōyamasenmaida bei ihrem ersten Feldtag.

図1 大山千枚田の棚田オーナーによる畑作第一日目。

Fig. 1 Rice field tenants, so-called *tanada ōnā* (*ōnā* being derived from the English 'owner'), in Ōyamasenmaida on their first working day.

Meinen Eltern gewidmet

Rahmenschrift Dissertation Pia R. Kieninger

Inhaltsverzeichnis

Zusammenfassung 1

概要 3

Abstract 5

1. Einleitung 7
1.1 Warum Japan? 8
1.2 *Satoyama* und das japanische Naturkonzept 8
1.3 Die Bedeutung des Reisbaus für die Biodiversität 14
1.4 Soziodemographischer Wandel als Herausforderung für Kulturlandschaft und Biodiversität 20
1.5 Lösungs-Modelle: Stadt-Land-Partnerschaften und das *Tanada* Ownership System zur Landschafts-und Biodiversitätserhaltung 22

2. Untersuchungsgebiet 26

3. Methodik 29

4. Kurzdarstellung der Artikel und ihrer Ergebnisse 31

5. Schlussfolgerungen und Ausblick 34

6. Literatur 35

7. Dank 40

8. Wissenschaftliche Publikationen 46

Artikel 1: Kieninger, P., Holzner, W., Kriechbaum, M. 2009. Biocultural Diversity and Satoyama. Emotions and the fun-factor in nature conservation – A lesson from Japan. Die Bodenkultur 60 (1): 15–21 ……..47

Artikel 2: Kieninger, P.R., Yamaji, E., Penker, M. 2011. Urban people as paddy farmers: The Japanese Tanada Ownership System discussed from a European perspective. Renewable Agriculture and Food Systems 26 (4): 328–341 ……………...…………………...57

Artikel 3: Kieninger, P.R., Penker, M., Yamaji, E. 2012. Esthetic and spiritual values motivating collective action for the conservation of traditional rural landscapes – A case study of rice terraces in Japan. Renewable Agriculture and Food System DOI: http://dx.doi.org/10.1017/S1742170512000269: 1–16 ……………..……………………………...………………………....……....73

Curriculum vitae, publications and presentation …………………………....…………….91

Rahmenschrift Dissertation Pia R. Kieninger

Zusammenfassung

Kulturlandschaftserhaltung durch urbanes Engagement in der traditionellen Landbewirtschaftung – Eine Fallstudie aus Japan

Biodiversität in der Kulturlandschaft ist abhängig von komplexen Wechselwirkungen zwischen ökologischen, ökonomischen und kulturellen Prozessen. Wie in vielen Teilen dieser Welt, so wird auch in Japan heutzutage der Schwund der traditionellen Kulturlandschaft und ihrer Biodiversität aufgrund sozioökonomischer Umwälzungen und dadurch bedingter Nutzungsaufgabe als ein Problem wahrgenommen. Ein vorwiegend naturwissenschaftlicher Ansatz im Naturschutz, wie er in Europa vorherrscht, mag dann kontraproduktiv sein, wenn er die Menschen aus der Natur ausschließt und verhindert, dass sie für 'ihre' Biodiversität Verantwortung übernehmen. Der Ansatz, der in Japan verfolgt wird, gründet in einem traditionellen Naturverständnis, das sehr stark von der Kultur geprägt ist und sich durch integrative und partizipative Strategien auszeichnet. In Japan gibt es nicht die starke Trennung zwischen berührter und unberührter Natur, die unser westliches Denken prägt. Die gestaltende Hand des Menschen in der Landschaft wird geradezu geschätzt und tendenziell werden Kulturlandschaften 'reiner Wildnis' vorgezogen. Der japanische Begriff *satoyama* ruft Emotionen hervor und umschreibt die Kulturlandschaft, die Einheit von Natur und Kultur, die daraus entstehende Biodiversität und ihre Bedeutung für unsere Lebensqualität. 'Basis-Elemente' von *satoyama* sind Dörfer am Fuß der Berge, sekundäres Grasland und Niederwälder auf den Hängen, sowie Reis und andere Ackerfrüchte in den Ebenen oder oft auch auf Terrassen angebaut. Reisterrassen (auf Japanisch *tanada*) nehmen dabei eine besondere Stellung ein. Sie sind nicht nur wichtig wegen ihrer Ökosystemleistungen für den Menschen (z.B. Verhinderung von Erosion und Überschwemmung, Wasserreinigung, Grundwasserschutz sowie Nahrungsmittelproduktion), sondern auch wichtige Biotope für eine Vielzahl von Pflanzen und Tieren. Außerdem werden sie aufgrund ihrer landschaftlichen Schönheit und kulturellen Wertes wegen wertgeschätzt und geliebt. Steigendes Bewusstsein für die Bedeutung von *satoyama* haben Ende der 1990er zur Entstehung einer Vielzahl von

Freiwilligen-Landwirtschafts-Programmen geführt. Die vorliegende Doktorarbeit beschäftigt sich in einer Fallstudie in Ōyamasenmaida (Präfektur Chiba, etwa 80 bis 100 km südöstlich von Tōkyō entfernt), mit dem 'Tanada Ownership System', in dem v.a. StädterInnen Reisterrassen pachten und diese unter der Anleitung der EigentümerInnen oder lokaler BäuerInnen und ExpertInnen bewirtschaften. Das Tanada-Ownership-System entstand 1992 auf der Insel Shikoku und hat sich seitdem landesweit ausgebreitet. Im Gegensatz zur eher individualistischen, zeitlich limitierten Teilnahme von Freiwilligen in europäischen Modellen verspricht das Tanada-Ownership-System eine langfristige Stadt-Land-Partnerschaft und so die Wertschätzung lokalen Wissens, der natürlichen Ressourcen und die Erhaltung der Reisterrassenlandschaften. Die Ergebnisse zeigen, dass die landschaftliche Schönheit das Hauptmotiv der PächterInnen ist. Auch die BesucherInnen kommen hauptsächlich der schönen Reisterrassen wegen nach Ōyamasenmaida. Die 'aktiven' PächterInnen unterscheiden sich jedoch von den 'passiven' BesucherInnen in einem gesteigerten ökologischen Interesse und einer stärkeren emotionalen Verbundenheit zur Landschaft. Der Glaube an eine beseelte Natur ('belief in nature spirits') ist bei beiden Gruppen überraschend hoch und ähnlich. Signifikante Unterschiede zwischen beiden Gruppen zeigen sich jedoch in der Kategorie derjenigen Personen, die angeben, dass sie immer daran glauben. Noch ausgeprägter ist dieser Unterschied, wenn man nur die Frauen betrachtet. Inwiefern dieser Glaube ein Motiv für die Teilnahme am Ownership-System ist, bzw. ob er erst durch die Teilnahme daran entstanden ist, konnte nicht eindeutig geklärt werden. Freiwillige verstärkt auch mit emotionalen und ästhetischen Motiven für den Landschafts- und Naturschutz zu mobilisieren, könnte auch für Europa eine vielversprechende Strategie sein.

Rahmenschrift Dissertation Pia R. Kieninger

概要
伝統的農業による文化的景観保存の市民活動 – 日本のケーススタディ

文化的景観の中の生物多様性は、生態的、経済的、文化的なプロセスの複雑な相互作用に依存している。伝統的な文化的景観とそこに含まれている生物多様性の社会経済的変化による劣化と、農業管理の放棄は、現在日本でも、世界の多くの国と同様に、問題として認識されている。ヨーロッパで一般的な自然科学に基づく自然保護のアプローチが、人間を自然から除外し、人々が「彼らの」景観に対する責任を担いにくくなるのであれば、逆効果かもしれない。

日本の自然保護は文化的影響の強い伝統的な自然概念に基づき、包括的な人間の手入れを強調する考えが特徴である。日本では、西洋の思考のような、人間の影響を受けた自然と原生的な自然のはっきりした区別はない。人間が手を加えた自然は尊重され、文化的景観は荒野の風景より好まれる傾向がある。「里山」の概念は感情を喚起し、文化的景観、自然と文化の統一、その結果である生物多様性と人間の生活の質に関連した意味を包括している。里山の「基本要素」は山のふもとにある村、低斜面の草原や二次林、平野や段丘にある田畑である。特に日本の棚田は重要な役割を果たしている。棚田は人間の暮らしに大事な役割を果たすほか（例えば、浸食や洪水の防止、水質浄化、地下水の涵養や食糧生産）、多様な動植物の重要なビオトープである。その上、棚田は景色の美しさと文化的価値によって評価され、愛されている。1990年代後半に里山の重要性に対する意識が高まり、様々なボランティアの農業プログラムが始まった。当論文では、大山千枚田（千葉県）のケーススタディにより、「棚田オーナー制度」を分析する。これは市民が棚田をリースし、棚田の所有者、地元の農家やプロの指導を受けながら農業を行う活動である。棚田オーナー制度は、1992年に四国で初めて行われ、以後、全国各

地に広まっている。欧州モデルでは、ボランティアは個々で期間限定の参加を行うのに対し、日本の棚田オーナー制度は、長期的な「都市と農村のパートナーシップ」に伴う地元の知識や自然資源の評価、または棚田風景の保全を振興させる。調査結果によると、オーナーの主な動機は風景の美しさであった。訪問者も主に美しい棚田を見るために大山千枚田に来る。しかし、「アクティブな」オーナーは「パッシブな」訪問者と比べて生態系に対する関心や景観への愛着が強い。自然の精霊については両グループで驚くほど広く信じられている。常にそれを信じている、と答えた人の割合は、オーナーの方が訪問者より高い。女性のみを考慮すると、この差はさらに大きい。この信念がどの程度棚田オーナー制度への参加の動機であるか、またはこの信念が参加を通じて強まったのかは、明らかにすることはできなかった。ヨーロッパでも、日本のような感情的・美的動機による風景・自然保全のボランティア活動は成功するかもしれない。

Abstract

Cultural landscape conservation through urban engagement in traditional land use management – A case study from Japan

In cultural landscapes, biodiversity depends on a complex interplay of ecological, economic and cultural processes. As in many parts of the world, so also in Japan, the degradation of the traditional cultural landscape and its biodiversity due to socio-economic changes and the abandonment of agricultural management is regarded as a problem. A predominantly scientific approach to nature conservation, as is prevalent in Europe, might be counterproductive if it banned humans from nature and discouraged them from taking responsibility for 'their own' biodiversity.

In Japan, traditionally, untouched nature is not distinguished from nature shaped by culture; the formative hand of man in nature is even appreciated. The Japanese term *satoyama* stands for traditional rural landscape, including its biodiversity and social traditions, and evokes emotions. 'Basic elements' of *satoyama* are villages at the foot of the mountains, secondary grasslands and coppiced forests on the slopes, rice and other arable crops grown in the plains or also often on terraces. Rice terraces (called *tanada*) play a special role. Apart from food production and their relevance for other ecosystem services, e.g. erosion and flood prevention, groundwater protection or water purification, they are important habitats for fauna and flora and are appreciated and loved as places of high cultural value. Increasing awareness of the importance of *satoyama* led in the 1990s to the creation of numerous voluntary urban farming programes. Through a mixed methods case study in Ōyamasenmaida (Chiba Prefecture, 80–100 km south-eastern from Tōkyō) this doctoral thesis focuses on a highly relevant rural–urban cooperation, known as the *Tanada* Ownership System, under which landowners lease out their rice terraces mainly to city dwellers for these to grow their own rice under the instruction and with the well-organized support of local farmers and other local experts. The first *Tanada* Ownership System started in 1992 on Shikoku Island and spread over the whole country. In contrast to the more short-term individualistic European models, the Ownership System promises long-term rural–urban relations, regard for local knowledge and natural resources, as well as the maintenance of the rice terrace landscapes. Among other things, the results show that the beauty of the landscape is the main motivation for the tenants to participate, as well as for visitors to come and enjoy the rice terrace scenery. The 'active' tenants, however, differ from the 'passive' visitors through their ecological interest in and emotional attachment to the area. The belief in nature spirits is surprisingly widespread among tenants and visitors. A higher percentage of interviewees able to imagine the constant presence of such spirits in nature was found among the tenants than among the visitors. Even more significant in this respect is the difference between female tenants and female visitors. To what extent the belief in nature spirits is a reason for participation in the *Tanada* Ownership System or a result thereof, could not be answered conclusively within the framework of this study. Triggering emotional and aesthetic motivations could also be a promising way for mobilising conservation volunteers in Europe.

1. Einleitung

Die vorliegende Doktorarbeit – eine kumulative Dissertation – besteht aus drei wissenschaftlichen Artikeln in referierten Fachzeitschriften, die sich mit dem Erhalt biokultureller Vielfalt in der Kulturlandschaft Japans beschäftigen. Ein Artikel analysiert, wie sich das japanische Konzept der Mensch-Natur-Beziehung (im Vergleich zu dem Europas) auf den Umgang mit der Natur auswirkt und was dies für Naturschutzaktivitäten bedeutet. In einer Fallstudie wird ein Reisterrassen-Pachtsystem – das *Tanada*-Ownership-System in Ōyamasenmaida auf der Bōsō-Halbinsel (Präfektur Chiba, etwa 80 bis 100 km südöstlich von Tōkyō entfernt) – untersucht und in zwei Artikeln herausgearbeitet, welche Motive dem Engagement der vor allem urbanen TeilnehmerInnen für den Erhalt der traditionellen Kulturlandschaft zu Grunde liegen und inwiefern ästhetische und spirituelle Werte dabei eine Rolle spielen.

Die nun folgenden Kapitel bilden einen Rahmen um diese drei Artikel. Im Kapitel 1.1 stelle ich dar, wie mich meine Dissertation nach Japan und dort zu meinem Untersuchungsgegenstand geführt hat. Galt mein Ausgangsinteresse der Biodiversität japanischer Reisterrassen, so weckte dann insbesondere das *Tanada*-Ownership-System – ein Reisterrassen-Pachtsystem zur Erhaltung von Kulturlandschaft und Biodiversität – mein Erkenntnisinteresse. Die folgenden beiden Kapitel 1.2 und 1.3 führen in *satoyama*, die japanische Kulturlandschaft, und in die Bedeutung des Reisbaus für die Biodiversität ein. Da Letzteres von den beiliegenden Artikeln nur am Rande thematisiert wird, widme ich der Biodiversität in dieser Rahmenschrift etwas mehr Raum (siehe Kapitel 1.3). Kapitel 1.4 erläutert, wie der sozio-demographische Wandel zur Nutzungsaufgabe führt und somit die Kulturlandschaft und ihre Biodiversität bedroht werden, bevor dann einige Lösungsansätze in Form von Stadt-Land-Partnerschaften, vor allem aber das *Tanada*-Ownership-System, als zentraler Untersuchungsgegenstand dieser Arbeit, vorgestellt werden (siehe Kapitel 1.5). Kapitel 2 stellt das Untersuchungsgebiet vor und Kapitel 3 den methodischen Zugang. Schließlich werden die Ergebnisse der drei beiliegenden Artikel kurz zusammengefasst und in Kapitel 5 wird ein kurzes Resümee gezogen sowie ein Ausblick auf künftige Forschungsfelder gegeben.

1.1 Warum Japan?

Nach Japan gelangte ich aus einer Mischung aus Zufall, Abenteuerlust und Neugier, ein mir bisher völlig unbekanntes, weit entfernt liegendes Land mit einer fremden Kultur kennenzulernen. Das ursprüngliche Thema meiner Doktorarbeit war eine vegetationsökologische Forschung zur Biodiversität einer Reisterrassenlandschaft in einer alten traditionellen Kulturlandschaft Japans, unter der Betreuung von Prof. Masahiko Ohsawa (Tōkyō University). Die Ergebnisse der Erhebungen zeigen erstaunliche Parallelen zur Entwicklung der Kulturlandschaft in Europa. In meinem Untersuchungsgebiet war die ursprüngliche Vielfalt an Lebensräumen und damit auch die Phytodiversität sehr stark durch eine Extensivierung der Nutzung bzw. ihre gänzliche Aufgabe bedroht. Die Ergebnisse dieser vegetationsökologischen Untersuchungen sind noch nicht publiziert.

Im Zuge dieser Erhebungen bin ich auf das *Tanada*-Ownership-System gestoßen, das mein Interesse geweckt hat. Der Ansatz, Kulturlandschaft mit Freiwilligen zu erhalten, begann mich zu faszinieren und ich startete Erhebungen unter der Anleitung von Prof. Eiji Yamaji (Tōkyō University), um die Motive der in diesem System beteiligten Personen zu erkunden. Die Ergebnisse dieser Untersuchung werden in den beigelegten drei Publikationen vorgelegt.

1.2 *Satoyama* und das japanische Naturkonzept

Satoyama ist ein alter Begriff – 1759 wurden damit 'Berglandschaften nahe von Dörfern' bezeichnet (TOKORO, M. 1980, ZITIERT IN TAKEUCHI 2003). *Satoyama* besteht aus zwei chinesischen Schriftzeichen: *sato* (里) Dorf und *yama* (山) Berg. In den frühen 1960ern wurde das Wort wiederentdeckt, leicht modifiziert zu 'Berge nahe dem Dorf' und später für landwirtschaftlich genutztes Waldland verwendet (TAKEUCHI 2003; SHIDEI, T. 2000, ZITIERT IN TAKEUCHI 2003). Im Zusammenhang mit der in den späten 1960ern einsetzenden, staatlich vorangetriebenen Modernisierung der Landwirtschaft und starken Urbanisierung weiter Teile der traditionellen Kulturlandschaft, die im wahrsten Sinne des Wortes plattgemacht wurden und verschwanden, wandelte sich der Begriff mit der steigenden Sensibilisierung der Bevölkerung zum Inbegriff für die (schwindende) traditionelle japanische Kulturlandschaft (TAKEUCHI 2003). Heute wird *satoyama* in verschiedenen Kontexten verwendet, v.a. aber, um damit Natur in der Kulturlandschaft zu bezeichnen (IBID.). Viel mehr als beim deutschen Begriff 'Kulturlandschaft' oder dem englischen 'cultural landscape', schwingen bei *satoyama* auch Emotionen mit. *Satoyama* ist Heimat, für Leute aus dem Dorf und der Stadt gleichermaßen. *Satoyama* steht für Tradition, ist das Inbild der japanischen Landschaft – eine Art japanisches Arkadien – und viele JapanerInnen sehen in *satoyama* den Rest der ursprünglichen und unberührten Naturlandschaft des Landes (KIENINGER ET AL. 2012; TAKEUCHI 2003).

Der Schutz von *Satoyama*-Landschaften als Hotspots von Biodiversität und biokultureller Vielfalt ist zu einem wichtigen gesellschaftlichen Anliegen und wissenschaftlichen Forschungsthema geworden (e.g. MORIMOTO 2011; IWATA ET AL. 2010; KOBORI 2009; MIYAURA 2009; OHSAWA & KITAZAWA 2009; YAMADA ET AL. 2007; TAKEDA ET AL. 2006; NATORI ET AL. 2005; KOBORI & PRIMACK 2003; TAKEUCHI 2003, 2001; FUJIOKA & YOSHIDA 2001; KATO 2001; WASHITANI 2001). *Satoyama* ist auch vorrangig in der japanischen Biodiversitätsstrategie politisch verankert (MOE 2010A). Die Food and Agriculture Organization of the United Nations (FAO) hat *Satoyama*-Landschaften in die Liste der 'Globally Important Agricultural Heritage Systems' aufgenommen (2008). 2008 gründete das Japanische Umweltministerium gemeinsam mit dem Institute for Advanced Studies der United Nation University, mit Unterstützung der UNESCO, die internationale Partnerschaft 'Satoyama Initiative', die 2010 auf der 10. Vertragsstaatenkonferenz der UN-Konvention über biologische Vielfalt offiziell in Start ging und weltweit den Schutz und die nachhaltige Nutzung menschlich geprägter Landschaften ('living in harmony with nature') zum Ziel hat (TAKEUCHI 2010; SATOYAMA INITIATIVE S.A.).

Satoyama ist eng mit dem Naturkonzept der JapanerInnen verbunden. Vom Menschen unberührte Naturlandschaft gibt es in Japan kaum (MOE 2010B). Wie schon das Wort *satoyama* den Menschen buchstäblich beinhaltet, so ist auch aus dem japanischen Naturverständnis, das seine Wurzeln in der shintoistischen und buddhistischen Kultur hat, der Mensch nicht wegzudenken: der Mensch als ein Teil der Natur (ISHII & NAKAMURA 2012; SAITO 2010; NAGASAWA 2008; KALLAND 1995); Mensch und Natur, vereint zu einem Ganzen (ISHII & NAKAMURA 2012; KALLAND 1995). Den ForscherInnen HAYASHI (2002) und ISHII & NAKAMURA (2012) gemäß, gab es in Japan über lange Zeit nicht einmal einen eigenen Ausdruck für Natur, losgelöst von 'Mensch'. Erst die Verwestlichung des Landes (seit der Meiji-Zeit, 1868 – 1912) brachte dies mit sich (MURASUGI, S. 1998, ZITIERT IN ISHII & NAKAMURA 2012).

Was aber bedeutet das? Sind die JapanerInnen damit die besseren NaturschützerInnen, da sie im Einklang mit der Natur leben? Ein Bild, das sich durch die japanische Literatur und die bildenden Künste bei uns im Westen als Stereotyp manifestiert hat (KALLAND 1995). Oder aber ist genau das Gegenteil der Fall? Dieser Gedanke kann dann aufkommen, wenn man sich die riesigen, lauten, energieintensiven Metropolen vor Augen führt, wenn man an den Raubbau Japans in den tropischen Wäldern anderer Länder denkt oder an das lautstark vom Ausland angeprangerte Thema 'Walfang'. Oder bedeutet es also, dass mit 'mir', als ein Teil der Natur, auch mein Handeln und Eingreifen natürlich sind? Und unterscheidet sich das dann wesentlich von dem 'Mach Dir die Erde untertan'-Prinzip der jüdisch-christlichen geprägten Welt[1]? Ich möchte nicht tiefer in diesen theologisch-philosophischen Diskurs einsteigen. Augenscheinlich und Gegenstand zahlreicher wissenschaftlicher Untersuchungen war und ist jedoch, dass JapanerInnen eine besondere Affinität für vom Menschen ästhetisch gestaltete und geformte Natur haben, während ihnen – allgemein gesprochen – echte Wildnis nicht ganz geheuer ist (SAITO 2002; KELLERT 1991). Und dies steht wiederum ganz im Gegensatz zum romantischen westeuropäisch-nordamerikanischen Traum unberührter, mystischer Natur (KOHSAKA ET AL. 2004; KELLERT 1993). Während dort der Mensch als

---

[1] Am bekanntesten dazu ist wohl Lynn Whites These der 'dominion over nature' (1967). Auch japanische AutorInnen greifen diesen Gedanken auf und erklären den veränderten japanischen Umgang mit der Natur durch den Einfluss des Westens ab der Meiji-Periode (z.B. OYADOMARI 1989; MUROTA 1985). Es gibt aber auch andere ForscherInnen, v.a. aus jüngerer Zeit, welche die These von White ablehnen und einen positiven Einfluss von Religion auf naturschutzrelevante Bereiche in ihren Studien nachweisen können (z.B. SHERKAT & ELLISON 2007; WOODRUM & WOLKOMIR 1997).

Eindringling und Bedrohung gesehen wird, haben JapanerInnen eine ausgeprägte Vorliebe für 'halb-natürliche Natur', in die der Mensch hineingehört, die er bewirtschaftet, perfektioniert und vollendet (KOHSAKA ET AL. 2004; KELLERT 1993, 1995). Und da sind wir wieder bei *satoyama* angelangt.

Abb. 2 *Satoyama*-Landschaft (Echigo-Tsumari) mit Reisterrassen und Niederwäldern. Ein Haus ist teilweise mit *Miscanthus* gedeckt.

図2 棚田と森の里山風景　(越後妻有)。茅葺屋根の一軒家。

Fig. 2 *Satoyama*-landscape (Echigo-Tsumari) with rice terraces and coppiced forests. One of the houses is partly thatched with *Miscanthus*.

Wie kann man sich *satoyama* aber nun konkret vorstellen? Typischerweise besteht *satoyama* aus fixen 'Grund-Elementen'. Zuerst wäre da einmal das Dorf (*sato*). Die Dörfer liegen oft am Fuß von Bergen (*yama*). Um die Häuser herum sind Obst- und Gemüsegärten angelegt. In den ebenen Flächen, nahe den Dörfern, werden Reis und andere Feldfrüchte (Buchweizen, Soja etc.) angebaut bzw. oft auch auf Terrassen. Bambushaine spielen eine wichtige Rolle für die Bereitstellung von Baumaterial, aber auch von Gemüse (Bambussprossen). Über die Berghänge erstrecken sich Niederwälder und die nur mehr marginal vorhandene 'japanische Heide' (*hara*). Die letzten Reste von Hochwald findet man meist im Umkreis von kulturell und religiös bedeutsamen Orten wie Schreinen und Tempeln. Durch die Nassfeldreiskultur bedingt, durchzieht ein weites und ausgeklügeltes System für Bewässerung und Wasserspeicherung die ganze Landschaft.

Wer in Japan schon einmal ein altes Bauernhaus betreten hat, wird wohl recht über das rußfarbene Innere gestaunt haben. In Japan gab es das System der Schornsteine nicht. Die Häuser waren einfach nur sehr hoch und der Rauch ging nach oben. Um den Rauch so gering wie möglich zu halten, wurde nur mit Holzkohle geheizt. Das Holz dazu stammte von den als Niederwäldern bewirtschafteten Berghängen (*yama*). Die japanischen Niederwälder sind den österreichischen in Aussehen und Artenzusammensetzung sehr ähnlich (HOLZNER 1983). Typisch sind sommergrüne Eichen (*Quercus serrata, acutissima, mongolica*) und *Castanea crenata* (IBID.). Die Umtriebszeiten liegen zwischen 7 bis 30 Jahren (IBID.). Das Laub der Niederwälder wurde als Einstreu für die Tiere im Stall und als Dünger für die Felder verwendet. Niederwälder waren auch wichtige Orte für die Pilzzucht, v.a. für den inzwischen auch in Europa bekannten *shiitake* (*Lentodes edulis*) (IBID.). Durch die hohen Lohnkosten in der Forstarbeit und v.a. durch das Aufkommen von Elektrizität und fossiler Energie nach dem Zweiten Weltkrieg (TAKEUCHI 2010) geht die Niederwaldbewirtschaftung jedoch immer mehr zurück. Das Ergebnis ist ein undurchdringbarer Unterwuchs, v.a. mit Bambus (*Sasa* sp. und *Pleioblastus chino*), sodass auch die Wildtiere (Wildschweine, Affen, Bären, *Shika*-Hirsche etc.) ihr ehemaliges Zuhause im Niederwald verlassen und auf der Suche nach Nahrung in die Dörfer kommen. Bisweilen umgeben hohe Zäune und andere Schutzeinrichtungen die Dörfer, um die Ernte zu schützen. Besonders schwierig ist dabei der Umgang mit Affen – zu denen die JapanerInnen eine Art Hassliebe verspüren (MARUYAMA 2006; SPRAGUE & IWASAKI 2006).

Samuraifilm-LiebhaberInnen wohlbekannt ist die *hara* – das endlose Meer von im Winde wogendem Hochgras (*Miscanthus* sp.) –, durch das hindurch der mutige Krieger auf dem

Weg zur Schlacht reitet, in der Ferne der schneebedeckte Berg Fuji. Dieses Bild gehört jedoch der Vergangenheit an, da von diesem wesentlichen Landschaftselement traditionell japanischer Kulturlandschaft, das ehemals weite Landesteile bedeckt hat, heute nur mehr ein marginaler Rest vorhanden ist. Bei uns wird *hara* gewöhnlich mit 'Heide' übersetzt, obwohl dieser Terminus im Grunde nicht korrekt ist (HOLZNER 1983). Das, was sie mit der europäischen Heide allenfalls verbindet, ist ihre Abhängigkeit von menschlicher Bewirtschaftung. *Hara* ist eine Hochgraswiese, mit dem Charaktergras *Miscanthus sinensis*, das 1 bis 2 m hoch werden kann (NUMATA 1974). Eine große Rolle spielen auch niedrige Gehölze, wie die Leguminosenstaude *Lespedeza bicolor* und *Rhododendron*-Arten. Ansonsten ist die japanische Heide mit etwa 25 Arten, im Vergleich zu europäischem Grasland, ziemlich artenarm (HOLZNER 1983). Von wirtschaftlichem Interesse war sie über lange Zeit zum einen wegen der Verwendung als Dachdeckmaterial, und zum anderen wurde *Miscanthus* verbrannt und die Asche als Dünger auf die Reisfelder gestreut.

Abb. 3 Mit *Miscanthus* gedecktes Hausdach.
図3　茅葺屋根。
Fig. 3 *Miscanthus* thatched roof.

Abb. 4 *Miscanthus* sp.
図4　ススキ。
Fig.4 *Miscanthus* sp.

Beginnend mit der Meiji-Periode (1868 – 1912), besonders dann aber ab den 1960ern (TAKEUCHI 2010) wurden Kunstdünger populär. Auch wurden die Häuser nicht mehr mit Stroh gedeckt (die wenigen Ausnahmen, die man heute noch sieht, sind Heimatmuseen). Mit dem Verlust der wirtschaftlichen Bedeutung wurde die *hara* nicht mehr genutzt und teilweise aufgeforstet (v.a. mit *Cryptomeria japonica* und *Chamaecyparis obtusa*).

Nun kann man sich fragen, ja aber, gibt es denn dann *satoyama* überhaupt noch? Das Japanische Umweltministerium definiert 40 % des Landes als *Satoyama*-Landschaft (MOE 2010B). Aber es ist ein anderes *satoyama*, nicht das von vor 100 oder 200 Jahren. Und überhaupt gab es nie 'die' *Satoyama*-Landschaft schlechthin. Natur und Landschaft sind nichts Stabiles. Sie durchleben einen ständigen Wandel, und genau das macht ihre Natürlichkeit aus. Ohne den Eingriff des Menschen in die Landschaft hätte *satoyama* einerseits gar nie entstehen können. Vor dem Einfluss des Menschen war Japan, bis auf wenige Ausnahmen (um die Vulkane herum oder in alpinen Zonen), komplett mit Wald bedeckt (HOLZNER 1983). Auch war es nicht so, dass der Mensch früher in einer kompletten Harmonie mit der Natur gelebt hätte (TAKEUCHI 2010; HOLZNER 1983). Heute ist nur der Eingriff, aufgrund der uns zur Verfügung stehenden Mittel, massiver (Biodiversitätskrise Nr. 1, siehe MOE 2010A). Andererseits aber ist die *Satoyama*-Landschaft heutzutage genau durch das Gegenteil, nämlich durch die Nutzungsaufgabe (Biodiversitätskrise Nr. 2, IBID.), bedroht. Aufgrund der soziokulturellen und ökologischen Bedeutung der *Satoyama*-Landschaft steht die japanische Gesellschaft vor der Herausforderung, traditionelle Landschaftsformen erhalten zu wollen, die durch – infolge der Industrialisierung – obsolet gewordene Wirtschafts- und Lebensformen entstanden sind und ohne diese verloren zu gehen drohen.

1.3 Die Bedeutung des Reisbaus für die Biodiversität

Durch die Nassfeld-Reisbaukultur und das dazugehörige Netzwerk von Bewässerung, Entwässerung und Wasserspeicherung ist ein weitverzweigtes, vielfältiges und einzigartiges aquatisches Ökosystem entstanden, das im Zusammenhang mit den angrenzenden terrestrischen Habitaten (Grasland, Brache, Wald[2]) eine große Vielfalt an unterschiedlichen Biotopen darstellt. Ein Lebensraum davon ist z.B. der etwa 1 bis 3 m

---

[2] 'Hinter jedem Reisfeld steht ein Wald': Um in der Edo-Ära ein Reisfeld von einem Hektar mit ausreichend Dünger versorgen zu können, waren mehrere Hektar Wald notwendig (KOBORI & PRIMACK 2003).

breite Graslandgürtel *kariage-ba* (Mähplatz), ein Ökoton zwischen Reisfeldern und Wald. Bis zur Industrialisierung der Landwirtschaft mähten die BäuerInnen diesen Bereich genauso regelmäßig wie die Böschungen zwischen den Terrassen, um die Beschattung ihrer Reisfelder durch aufkommendes Gebüsch zu vermeiden (OHSAWA & KITAZAWA 2009; KITAZAWA & OHSAWA 2002). Das regelmäßige Mähen trug dazu bei, dass speziell angepasste Graslandbewohner, welche sonst in der japanischen Landschaft keine Lebensmöglichkeiten gehabt hätten, bis heute erhalten geblieben sind. Daher ist dieses Biotop sehr artenreich, insbesondere auch an vielen seltenen Arten (wie *Gentiana scabra* und *Campanula punctata*). Nach einer Hypothese von Prof. Masahiko Ohsawa (mündliche Mitteilung 2004) gehören dazu auch Steppenpflanzen (wie *Campanula cirsium*), die während der Eiszeit, als Japan durch eine Eisbrücke mit dem asiatischen Kontinent verbunden war, eingewandert waren. Durch Hilfe der regelmäßig mähenden BäuerInnen konnten sie bis heute in der *kariage-ba* überdauern. Sind diese Steppenpflanzen einmal ausgestorben, ist es sehr schwer möglich, solche Arten wieder zurückzubringen (KITAZAWA & OHSAWA 2002).

Die Habitat-Vielfalt ist der Grund für die ungeheure Vielfalt von Arten. Einer Studie zufolge, die auf der 10. Vertragsstaatenkonferenz der UN-Konvention über biologische Vielfalt vorgestellt wurde, sind in Japan 5.668 Tier- und 2.075 Pflanzenarten an das Ökosystem Reisfeld angepasst (IWABUCHI ET AL. 2010).

Abb. 5 Kaulquappen im Frühjahr in der Reisterrasse.

Abb. 6 Laubfrosch im Schilf derReisterrassen-böschung

図5　春の棚田の中のおたまじゃくし。

図6　棚田脇のアシに乗った日本雨蛙。

Fig. 5 Tadpoles in the paddy fields in spring.

Fig. 6 Tree frog on a reed on a rice terrace slope.

Etwa die Hälfte aller Rote-Liste-Arten Japans hat heute in der *Satoyama*-Landschaft ihren Lebensraum (ISHII & NAKAMURA 2012). Der Grund dafür ist, dass 60 % der natürlichen Feuchtgebiete (z.B. Auen), die es in der Meiji- (1868 – 1912) und in der Taisho- (1912 – 1926) Ära noch gab, heute, v.a. aufgrund der landwirtschaftlichen Neulandgewinnung der Nachkriegsjahre, nicht mehr vorhanden sind (WASHITANI 2007; GEOGRAPHICAL SURVEY INSTITUTE 2000). Reisfelder, als 'agrarische Feuchtgebiete' (KOBORI 2009), stellen daher die größte Feuchtgebietsfläche des Landes (die Hälfte aller Süßwasser-Feuchtgebiete) dar (KOBORI & PRIMACK 2003). Sie sind damit einerseits zum Ersatzbiotop für viele Tier- und Pflanzenarten geworden, die seichte, nährstoffreiche Gewässer mit saisonalen Schwankungen als Lebensraum benötigen (WASHITANI 2007). Viele Tiere und Pflanzen haben sich darüber hinaus, über Jahrtausende, an den Jahreszyklus des Reisbaus – den Wechsel zwischen nass und trocken – angepasst.

So nutzen manche Tiere etwa bestimmte Bereiche nur in bestimmten Phasen ihres Lebens oder nur zu ganz bestimmten Jahreszeiten, wie zum Beispiel zahlreiche Froscharten. *Rana japonica* und *Rhacophorus* spp. leben im Wald und nutzen die im Frühjahr angrenzenden, wassergefüllten Reisfelder zum Laichen (NATUHARA 2012). Darüber freuen sich fröschejagende Schlangen und der Graugesichtsbussard (*Butastur indicus*).

Vor der intensiven Modernisierung der Landwirtschaft in den 1960ern bis 1980ern waren die Reisfelder über Bewässerungsgräben/-kanäle mit Bächen, Flüssen und Seen verbunden. Das Verbundsystem Fluss/See – Bewässerungsgraben/-kanal – Reisfeld ist eine wichtige Nahrungsroute für viele Mollusken und Fische. Im Frühjahr lockt das bereits lauwarme Wasser der Reisfelder die Fische aus den Flüssen und Seen (z.B. Karpfen, aber auch viele andere kleinere Fische), gegen den Strom zu schwimmen, um in den wärmeren Gefilden der Reisfelder abzulaichen (IBID.). Aber auch für viele andere Tiere ist das Ökosystem Reisfeld ein wichtiger Ort für Reproduktion und Nahrung.

So hat das *Heike*-Glühwürmchen (*Luciola lateralis*) seinen Lebenszyklus komplett auf den Reisbau eingestellt. Von Juni bis August legt das Weibchen Eier auf Moose oder die feuchte Oberfläche von Pflanzen, die seitlich an den Wänden der Bewässerungsgräben oder am Rand der Reisterrassen wachsen (KOJI ET AL. 2012). Die geschlüpften Larven wandern ins Wasser, wo sie sich ausschließlich von Schnecken ernähren. Bevor sie im Schlamm am Grund des Reisfeldes überwintern, machen sie vier Larvenstadien durch (IBID.). Im Frühling fressen sie sich nochmal kräftig mit Schnecken voll, häuten sich und krabbeln dann an Land, wo sie sich in der Erde der Reisfeldböschungen verpuppen (IBID.). Im Juni schlüpfen die neuen erwachsenen Tiere. Diese Lebensphase, in der sie ohne jegliche Nahrung auskommen, ist nur sehr kurz. Nachdem der/die PartnerIn gefunden wurde, kommt es zur Paarung in der niederen Vegetation am Rand der Reisterrassen und eine neue Glühwürmchen-Generation startet ins Leben (IBID.). Auch das berühmteste (weil besonders groß und hell leuchtend) Glühwürmchen des Landes, das *Genji-botaru* (*Luciola cruciata*), lebt bisweilen als Larve in den Bewässerungsgräben der Reisfelder (TAKEDA ET AL. 2006). Ein weiteres Beispiel für ein Insekt, das sein Leben auf das traditionelle Reisbau-Ökosystem angepasst hat, ist die Libelle (WASHITANI 2001). In Japan gibt es um die 180 bis 200 Libellen-Arten (die Zahl schwankt je nach AutorIn: 180 Arten bei PRIMACK ET AL. 2000; 190 Arten bei KONISHI 2004; 200 Arten bei WASHITANI 2001). Besonders dekorativ ist die endemische *Sympetrum frequens* – auf Japanisch, ihrer auffallend roten Farbe wegen, *akatombo* (Rote Libelle) oder *akiakane* (Rotgefärbter Herbst) genannt. Ihr drastischer Rückgang ist auf den intensiven Einsatz von Insektiziden im Reisbau zurückzuführen (UEDA, T. 2007, zitiert in ISHII & NAKAMURA 2012). Das betrifft die Glühwürmchen gleichermaßen. Bereits 1924 und 1925 wurden in den Präfekturen Shiga und Nagano Schutzgebiete für das Glühwürmchen ausgewiesen, allerdings mit mäßigem Erfolg, da ihr Rückgang auch durch den Schwund ihres Lebensraumes verursacht ist (ISHII & NAKAMURA 2012).

Wie das Glühwürmchen, so hat auch die Libelle eine hohe kulturelle Bedeutung in Japan: Bereits in Funden aus der Yayoi-Zeit (2. – 3. Jahrhundert n. Chr.) fand man Abbildungen von Libellen, gemeinsam mit Spinnen und Gottesanbeterinnen, allesamt Räuber von Ernteschädlingen, als Verzierung auf zeremoniellen Glocken, was von HistorikerInnen als Gebet um eine gute Ernte interpretiert wird. Das ist nur ein Beispiel dafür, wie eng der Reisbau mit der japanischen Kultur verbunden ist (KONISHI 2004).

Da die Libellen als *kachi-mushi* (Sieger-Insekten) angesehen wurden, verzierten die Samurai mit ihnen ihre Helme (IBID.). Libellen und Glühwürmchen, aber auch viele andere Insekten – den JapanerInnen sagt man nach, dass sie die größten Insektenfans der Welt sind (IBID.) – spielen seit alters her eine bedeutende Rolle in der japanischen Literatur und bildenden Kunst: *„The familiar cultural relationship between the Japanese and insects can be found in pictures and literature, such as folktales, Waka poems and Haiku poetry. There has been a culture of insect-listening parties from old times, and insect sellers selling crickets were present in the Edo period"* (ISHII, M. 2003, zitiert in ISHII & NAKAMURA 2012, S. 341). Insektenfangen ist bis heute ein beliebtes Kinderspiel und in den lauen Sommernächten begibt man sich zum gemeinsamen *hotaru-mi*: Glühwürmchenschauen. Das *Genji*-Glühwürmchen zählt in mindestens 10 Bezirken zum nationalen Kulturerbe (TAKEDA ET AL. 2006).

Reisfeldlandschaften sind auch wichtige Nahrungs-, Nist- und Rastorte für 30 % (135 Arten) der 430 einheimischen Vogelarten Japans und Koreas, von denen 32 Arten als national oder international gefährdet gelten (FUJIOKA ET AL. 2010). Zugvögel, wie zum Beispiel Gänse, Schwäne, Enten und Kraniche, verbringen den Winter über in Japan, wo sie sich auf den abgeernteten Reisfeldern von den zurückgelassenen Resten ernähren (FUJIOKA & YOSHIKA 2001). So auch die Blässgans (*Anser albifrons frontalis*), die den Sommer über in der 4.000 km entfernten russischen Tundra zubringt (JAPANESE ASSOCIATION FOR WILD GEESE PROTECTION 2005). Seit 1971 steht sie unter offiziellem Schutz, da sie so drastisch zurückgegangen ist (IBID.). Heute konnte sich ihre Population auf 60.000 erholen (IBID.)[3].

Zwei besondere Symbolvögel des Landes sind der Nippon-Ibis (*Nipponia nippon*, auf Japanisch *toki*) und der Orientalische Weißstorch, auch Schwarzschnabelstorch genannt (*Ciconia boyciana*, auf Japanisch *kōnotori*). Beide Vögel fallen, gemäß der Einteilung von

---

[3] Ihr Rückgang, wie auch der der Saatgänse (*Anser fabalis serrirostris, A. fabalis middendorffii*), der kleinen Schneegans (*A. caerulescens caerulescens*) und des Lilford-, Mönchs- und Weißnackenkranichs (*Grus grus lilfordi, G. monacha, G. vipio*), ist neben der Zerstörung ihres Lebensraumes auch auf die intensive Jagdtätigkeit in der Meiji-Zeit und in den 1970ern zurückzuführen (FUJIOKA & YOSHIKA 2001). Durch verbesserten Schutz (Jagdverbot) konnte in den letzten 40 Jahren ein Anstieg beobachtet werden (IBID.).

FUJIOKA und YOSHIKA (2001), in die Gruppe der 'Crop-nondependent farmland birds'. Das heißt, anders als die zuvor erwähnten Blässgänse, die sich als 'Crop-eating farmland birds' von Feldfrüchten (in diesem Fall Reis) ernähren, sind die Hauptmahlzeit von *toki* und *kōnotori* v.a. Fische, Insekten (z.B. Wasserkäfer), Amphibien (z.B. Frösche, Laiche) und Mollusken, also alles Tiere, die in einem gesunden Reisökosystem zu Hause sind. Die Ursachen, die zum Aussterben[4] des Weißstorchs und Nippon-Ibisses geführt haben, liegen im automatisierten, intensiven modernen Reisbau, der zu einer Fragmentierung der ehemals zusammenhängenden Ökosysteme geführt hat. Immer weniger Fischen und aquatischen Tierchen gelingt der Weg durch die betonierten Bewässerungskanäle, Drainagesysteme, Wehre und Schleusen in die Reisfelder hinauf. Der Einsatz von Pestiziden und chemischen Düngemitteln, die kurze Verweildauer von Wasser in den Feldern, Konsolidierungsmaßnahmen und das Ausräumen des Agrarlandes reduzierten zudem ihr Nahrungsangebot.

Um diese negative Entwicklung wenigstens teilweise wieder rückgängig zu machen, wird seit etwa den letzten zehn Jahren die *fuyu-mizu-tanbo* (Winter-Wasser-Reisfeld)-Methode[5] immer populärer, die bereits in der Edo-Periode bekannt war, aber in Vergessenheit geraten ist (MOE 2010C; JAPANESE ASSOCIATION FOR WILD GEESE PROTECTION 2005). So wird sie zum Beispiel in der Umgebung der Stadt Toyooka (Präfektur Hyōgo) angewandt, die dem Weißstorch wieder ein Zuhause geben will. 2011 lebten dort wieder 43 Individuen in freier Wildbahn (DAILY YOMIURI ONLINE 2011).

Bei der Winter-Wasser-Reisfeld-Methode sind die Felder einige Wintermonate hindurch mit Wasser bedeckt (im konventionellen Reisbau bleibt das Reisfeld, nach der Ernte im Herbst, bis zum nächsten Frühjahr trocken). Das fördert Tubifex-Würmer, eine wichtige Nahrung für andere Tiere (JAPANESE ASSOCIATION FOR WILD GEESE PROTECTION

---

[4] Den letzten Orientalischen Weißstorch in Wildnis sah man 1971 in Toyooka (Präfektur Hyōgo), die letzten Nippon-Ibisse in Freiheit starben 1981 aus (MIE 2012; MOE 2010D). Für beide Vögel wurden seit Mitte der 1960er aktive Zuchtprogramme und Wiederansiedelungen in Toyooka (Weißstorch) und auf der Insel Sado (Nippon-Ibis) betrieben (MOE 2010D, E). 2008 wurden die ersten Nippon-Ibisse aus der Gefangenschaft entlassen; diesen Frühling schlüpften die ersten drei Küken in Freiheit (MIE 2012).
[5] 1998 wurde die *fuyu-mizu-tanbo*-Methode von BäuerInnen des Shimpo-Bezirks (Präfektur Miyagi) wieder ins Leben gerufen, zum Schutze des 150 ha großen Feuchtgebietes Kabukuri-numa – Heimat von mehr als 220 (darunter 127 gefährdeten) Vogelarten (MOE 2010C).

2005). Die Fäkalien der Würmer legen sich als dünne Schicht über den Grund und verhindern Unkrautaufkommen (MOE 2010D). Insekten und Amphibien, wichtige Dezimierer von Ernteschädlingen, haben bei dieser traditionellen Reisbaumethode länger Zeit, sich zu entwickeln (IBID.). Zug- und Wasservögel finden den Winter über Nahrung und düngen ihrerseits, mit ihren Ausscheidungen, die Felder (JAPANESE ASSOCIATION FOR WILD GEESE PROTECTION 2005). Auch das aufkommende Algenwachstum fördert die Fruchtbarkeit des Bodens (IBID.).

Mit diesen Ausführungen wollte ich veranschaulichen, wie essenziell die traditionelle Bewirtschaftung der Reislandschaften gerade für die Biodiversität Japans ist. Im nächsten Kapitel werde ich auf die Ursachen eingehen, warum dieses Agrar-Ökosystem und die damit verbundene Biodiversität heutzutage hochgradig gefährdet sind und komme anschließend zu beispielhaften Lösungsansätzen und zum Fokus meiner Arbeit (1.5): dem *Tanada*-Ownership-System.

1.4 Soziodemographischer Wandel als Herausforderung für Kulturlandschaft und Biodiversität

Japan durchlebt seit Jahrzehnten einen starken soziodemographischen Wandel. Das Land altert. Derzeit ist etwa ein Viertel der japanischen Bevölkerung mindestens 65 Jahre, Tendenz steigend: 30 % sind die Prognosen für 2020 und 40 % für das Jahr 2050 (MIC 2010). Besonders hart trifft es dabei das ländliche Japan. Die jungen Leute wandern alle in die Städte ab. Zurück am Land bleiben die Alten. Diese Entwicklung stellt Japan nicht nur vor eine große soziale und ökonomische Herausforderung, auch kulturell und ökologisch gesehen, ist dieser Trend besorgniserregend, da der Erhalt der jahrtausendealten Kulturlandschaft und ihre biokulturelle Vielfalt (Pflanzen, Tiere, Traditionen, Bräuche etc.) von der Aufrechterhaltung der Bewirtschaftung abhängen (siehe Kapitel 1.3).

Abb.7 Dorfbevölkerung im Bezirk Echigo-Tsumari (Präfektur Niigata) beim Nachmittagstratsch.
図7 午後の村人の井戸端会議（越後妻有、新潟県）。
Fig. 7 Villagers in the Echigo-Tsumari area (Niigata Prefecture) having an afternoon chat.

Japan ist ein sehr bergiges und waldreiches Land. Etwa drei Viertel des Landes sind gebirgig und zwei Drittel sind bewaldet. An diese geographischen Gegebenheiten haben sich die Menschen über Jahrtausende hinweg angepasst und als Resultat ist eine sehr kleinteilige Kulturlandschaft entstanden. Die durchschnittliche Betriebsfläche ist, obwohl seit 2002 um 0,15 ha etwas angestiegen, im internationalen Vergleich mit 1,38 ha dennoch sehr gering und trägt auch dazu bei, dass die Produktionskosten international nicht wettbewerbsfähig sind (MAFF 2008A,B). Immer weniger Leute wollen und können daher allein von der Landwirtschaft leben. Folglich ist die landwirtschaftliche Fläche seit 1961 um 24 % auf 12 % der Landesfläche geschrumpft (MAFF 2008A). Die landwirtschaftliche Produktion liegt auf den Schultern von PensionistInnen: 60 % der Personen, die in der Landwirtschaft aktiv tätig sind (2,99 Millionen), sind 65 Jahre oder älter – das Durchschnittsalter liegt bei 65,8 Jahren (MAFF 2011, 2008A).

Obwohl das Landwirtschaftsministerium für 2009 67.000 'landwirtschaftliche' NeueinsteigerInnen verbuchen konnte (davon 15.000 unter 39 Jahre), bleibt zweifelhaft, ob damit, angesichts der demographischen und ökonomischen Situation vieler Betriebe, die in naher Zukunft benötigten HofnachfolgerInnen aufzutreiben sind (MAFF 2011).

Bei einer vom Landwirtschaftsministerium durchgeführten Studie gaben 25,1 % aller untersuchten landwirtschaftlichen Haushalte an, keine/n NachfolgerIn zu besitzen; bei 42,6 % war die Hofnachfolge unsicher bzw. sehr fragwürdig und nur bei 32,3 % war eine Übernahme fix (MAFF 2011).

Abb. 8 – 11   In der Landwirtschaft sind in Japan zum Großteil RentnerInnen aktiv.
図 8 – 11   ほとんどの農業者は年金生活者。
Fig. 8 – 11   In Japan most of those involved in agricultural management are retired.

1.5 Lösungs-Modelle: Stadt-Land-Partnerschaften und das *Tanada*-Ownership-System zur Landschafts- und Biodiversitätserhaltung

Angesichts der zunehmenden Nutzungsaufgabe agrarischer Flächen und der damit einhergehenden unerwünschten Landschaftsveränderungen haben sich in den letzten

Jahren, in unterschiedlichen Städten Japans, verschiedene Modelle entwickelt, um den drohenden Landschaftswandel und Biodiversitätsverlust in den angrenzenden ländlichen Regionen hintanzuhalten. Ich stelle zunächst beispielhaft zwei Aktivitäten in Reislandschaften vor, um dann auf den Untersuchungsgegenstand meiner Arbeit, das *Tanada*-Ownership-System, im Detail einzugehen.

Toyooka, wo Stadt, Präfektur und die japanische Landwirtschaftskooperative (JA) gemeinsam Werbung für die 'Weißstorch-freundliche-Landbewirtschaftungsmethode' (= die Winter-Wasser-Reisfeld-Methode) betrieben haben, ist ein schönes Beispiel dafür, wie eine aus heutiger wirtschaftlicher Sicht unrentable traditionelle Bewirtschaftungsweise auch ökonomisch sehr erfolgreich sein kann (MOE 2010D). Pro Hektar unterstützt die Stadt die BäuerInnen mit 7.000 Yen (etwa 73 Euro, Stand 31.07.2012), wenn sie ihren Pestizideinsatz auf mindestens 75 % und den von chemischen Düngemitteln auf 100 % reduzieren, ihre Felder tiefer und früher im Frühjahr fluten und auch im Winter nochmals für mindestens einen Monat lang Wasser in die Felder lassen (IBID.). 2009 wurden 7 % der Reisflächen (212,3 ha) so bewirtschaftet. Der Reisertrag liegt zwar ein Viertel unter dem des heute üblichen Anbaus. Dafür aber lässt sich der Reis deutlich (23–71 %) teurer verkaufen (IBID.). Die Weißstörche von Toyooka sind zu einem Label geworden, der Ökotourismus boomt und ist eine wichtige Einnahmequelle für die ganze Umgebung.

Die Erfolgsgeschichte des Nippon-Ibisses auf der Insel Sado (Präfektur Niigata) erzählt sich ähnlich. Sado war bekannt für seinen ausgezeichneten Reis, den *sado koshihikari*, der sich zu Höchstpreisen verkaufen ließ (MOE 2010E). Daher waren die BäuerInnen an biodiversitätsförderndem Reisbau nicht sonderlich interessiert (IBID.). 2004 ruinierte jedoch ein schlimmer Taifun die gesamte Ernte und der *sado koshihikari* verlor seinen Status am Reismarkt (IBID.). Aus Angst, dass dies in der Folge eine große Anzahl von LandwirtInnen dazu veranlassen könnte, mit der Landwirtschaft dauerhaft aufzuhören, startete die Stadt Sado 2008 das Programm *toki-to-kurasu-sato* (Aufbau von Dörfern in

Koexistenz mit dem Nippon-Ibis) (IBID.). 2010 wurden auf Sado 1.200 ha mittels der *fuyu-mizu-tanbo* (Winter-Wasser-Reisfeld)-Methode bewirtschaftet; angestrebtes Ziel ist das Ansteigen der Nippon-Ibis-Population auf 60 Individuen bis 2015 (IBID.).

Um das Verständnis für den Wert von Biodiversität und seinen Schutz zu fördern, rief die Stadt Sado außerdem jeden 2. Sonntag im Juni und jeden 1. Sonntag im August zum 'Rice Paddy Organism Survey Day' aus, an denen sich in der gesamten Region BäuerInnen und die interessierte Bevölkerung gemeinsam in den Reisfeldern auf Tier- und Pflanzensuche begeben (IBID.). Zudem wird von jedem verkauften Kilogramm zertifiziertem *toki-to-kurasu-sato*-Reis ein Yen für den Schutz des Vogels verwendet (IBID.). Biodiversität vermittelndes 'Pflanzen- und Tiere-Schauen' gibt es auch in anderen Präfekturen. In der Stadt Takashima (Präfektur Shiga) dürfen sich BäuerInnen drei Pflanzen, welche in ihren Reisfeldern wachsen und auf die sie besonders stolz sind, auswählen und bekommen dafür ein Zertifikat, mittels dem sie ihren Reis teurer verkaufen können (NATUHARA 2012). In Fukuoka bekommen LandwirtInnen diese Zertifikation für ein Biodiversitätsmonitoring auf ihren eigenen Feldern (IBID.).

Als besonders populär und erfolgreich hat sich das *Tanada*-Ownership-System erwiesen, das erstmals 1992 auf der Insel Shikoku in Yusuhara (Präfektur Kochi) startete und sich seitdem im ganzen Land ausgebreitet hat (NATIONAL FEDERATION OF LAND IMPROVEMENT ASSOCIATION 2012). Das Ownership-System ist eine Art Pacht-System für Nicht-LandwirtInnen, die verschiedenste Feldfrüchte anbauen. Besonders beliebt ist Reis auf Terrassen, aufgrund ihrer landschaftlichen Schönheit und Attraktivität und des stark kulturell-traditionellen Wertes, den Reis für die japanische Gesellschaft innehat. Ein weiterer (praktischer) Grund ist die 'handliche' Größe der Reisterrassen, die auch – für die meist landwirtschaftlich unerfahrenen städtischen PächterInnen – manuell bewirtschaftbar ist[6] (IG[7]: DIREKTOR, 18. JUNI 2005). T*anada* ist das japanische Wort für

---

[6] Die Größe der Reisterrassen in Ōyamasenmaida variiert zwischen 20 bis 900 m² (ŌYAMASENMAIDA CULTURAL LANDSCAPE PRESERVATION COMMITTEE 2006). Die größeren Terrassen bestellen Gruppen, kleinere Terrassen werden von Einzelpersonen und Familien bewirtschaftet.

Reisterrasse. Aufgrund der Anziehungskraft der Reisterrassen – für viele JapanerInnen sind sie das Sinnbild von Heimat – ist das *Tanada*-Ownership-System daher eine gute 'Bühne', um Leute aus der Stadt und vom Land zusammenzubringen (AGENCY FOR CULTURAL AFFAIRS 2003A). Dieser Stadt-Land-Austausch soll Freundschaften, das Bewusstsein für die Problematik der ländlichen Regionen und die Wichtigkeit des Erhalts von Kulturlandschaft schaffen und wird nicht nur von politischer Seite als sehr wichtig angesehen und gefördert (AGENCY FOR CULTURAL AFFAIRS 2003A). So sieht der Direktor des *Tanada*-Ownership-Systems von Ōyamasenmaida den Austausch von StädterInnen und der ländlichen Bevölkerung mindestens als gleichermaßen wichtig an wie die Bewirtschaftung der Reisterrassen selbst: *„Der Großteil der japanischen Bevölkerung lebt in Städten. Die Regierung ist in der Stadt. Große und wichtige Entscheidungen werden von StädterInnen in Städten gefällt. StädterInnen können die Regierung ändern und sie verdienen Geld. Geld, das notwendig ist, die Bergregionen zu erhalten."* (EI[7]: DIREKTOR, 18. JUNI 2005).

Abb. 12 & 13   Reisfeldpächter von Ōyamasenmaida.
図 12 & 13   大山千枚田の棚田オーナー。
Fig. 12 & 13   Ōyamasenmaida rice terrace tenants.

Da das Pachten landwirtschaftlichen Grundes in Japan Nicht-LandwirtInnen nicht gestattet ist, wurde in Ōyamasenmaida, meiner Modellregion, eine Zeit lang über eine Gesetzeslücke agiert, indem die Stadt Kamogawa als Zwischenpächterin auftrat und das Land gegen eine Gebühr an die PächterInnen weiterverpachtete (IG: DIREKTOR, 26. SEPTEMBER 2009; EI: DIREKTOR, 18. JUNI 2005). 2002 bis 2003 hat die japanische

---

[7] IG = informelles Gespräch, EI = Experten-Interview, TB = teilnehmende Beobachtung, Direktor = Direktor vom Ōyamasenmaida-Ownership-System.

Regierung unter Koizumi Junichirō jedoch, um die Wirtschaft, die Revitalisierung der ländlichen Regionen und die Dezentralisierung zu fördern, die Ausweisung spezieller Sonderzonen bzw. Sonderbezirke (*sogo tokku*) zur Strukturreform beschlossen, in denen es u.a. Steuervergünstigungen gibt und bestimmte hemmende Gesetze dereguliert werden können (BOCHORODYCZ 2010; SALVINI ET AL. 2010; YUGAMI 2008). Die Sonderbezirke können für verschiedene Bereiche wie Wirtschaft, Bildung oder Landwirtschaft ausgewiesen werden, z.B. 'Farming village and city exchange', die Bewirtschaftung von Reisterrassen, das Brauen unfermentierten Reisweines etc. (YUGAMI 2008). Lokale Behörden (Städte/Gemeinden) müssen dazu einen Antrag stellen, bestimmte Gesetze zur Verfolgung konkret beschriebener Ziele aufzuheben[8] (YUGAMI 2008).

Die Stadt Kamogawa hat 2003 für Ōyamasenmaida einen Antrag als Reisterrassen-Landwirtschaft-*tokku* gestellt, der 2004 akzeptiert wurde (IG: DIREKTOR, 28. JÄNNER 2010; EI: DIREKTOR, 18. JUNI 2005). Seitdem können die LandwirtInnen in Ōyamasenmaida ihr Land direkt an Nicht-LandwirtInnen verpachten (YAMAJI 2006; KONAMI 2003), was jedoch die NPO 'Ōyamasenmaida-Schutzvereinigung' für die EigentümerInnen übernimmt (EI: DIREKTOR, 18. JUNI 2005). 10 % der Pachteinnahmen gehen an die EigentümerInnen, der Rest an die NPO (IBID.).

2. Untersuchungsgebiet

Das Untersuchungsgebiet der Fallstudie für diese Dissertation liegt etwa 80 bis 100 km von Tōkyō entfernt, am südlichen Zipfel der Bōsō-Halbinsel, inmitten der 'Berge'. Höchster Berg ist der ungefähr 219 m hohe Ōyama, der der Gegend, mit den sechs Gemeinden Hiratsuka, Narabayashi, Kamanuma, Sano, Kotsuka und Kobata, um seinen

---

[8] Welche Gesetze auf Antrag hin potenziell dereguliert werden können, wurde im Vorfeld in Zusammenarbeit von nationaler und regionaler Regierung, lokalen Verwaltungseinheiten, Universitäten, privaten Firmen etc. festgelegt (YUGAMI 2008). Das heißt, es kann nicht jedes beliebige Gesetz außer Kraft gesetzt werden, sondern der Antrag muss sich auf eine vorgegebene Liste an Gesetzen beziehen (IBID.). Die Zahl der *tokku*-Bezirke steigt beständig an (IBID.). Durch die Erdbebenkatastrophe in Tohoku im März 2011 hat die Regierung für diese Region die Etablierung zusätzlicher *fukko tokku* (reconstruction special zones) beschlossen (JAPAN TIMES 2012).

Fuß herum gelegen, ehemals ihren Namen gab (IG: PÄCHTERIN F, 10/2005; KAMOGAWA CITY 2004). Auf seinem Gipfel befindet sich ein Schrein-Tempel-Komplex (mit dem Takakura-Schrein und Ōyama-*fudōso*-Tempel). Diese Form von Synkretismus kommt auf dem Lande oft vor (SOOD & NASU 1995). 'Ōyamasenmaida' bedeutet übersetzt 'Großer [*ō*] Berg [*yama*] 1.000 [*sen*] Reisterrassen [*maida*]' und ist der Name für eine Reislandschaft, die aus vielen kleinen, hunderten von Reisterrassen besteht, die sich auf einen Höhenunterschied von 70 m erstrecken (siehe Abb. 14, große weiße Umrandung).

Die Reisterrassen sind in eine Landschaft aus kleinen Siedlungen, Feldern, Gemüse- und Obstgärten (v.a. Kaki und Sommerorange), Aufforstungen (v.a. *Cryptomeria japonica* und *Chamaecyparis obtusa*), Niederwäldern (z.B. *Quercus serrata*, *Quercus glauca*, *Castanopsis cuspidata*), Grasland und Bambushainen eingebettet (ŌYAMASENMAIDA CULTURAL LANDSCAPE PRESERVATION COMMITTEE 2006).

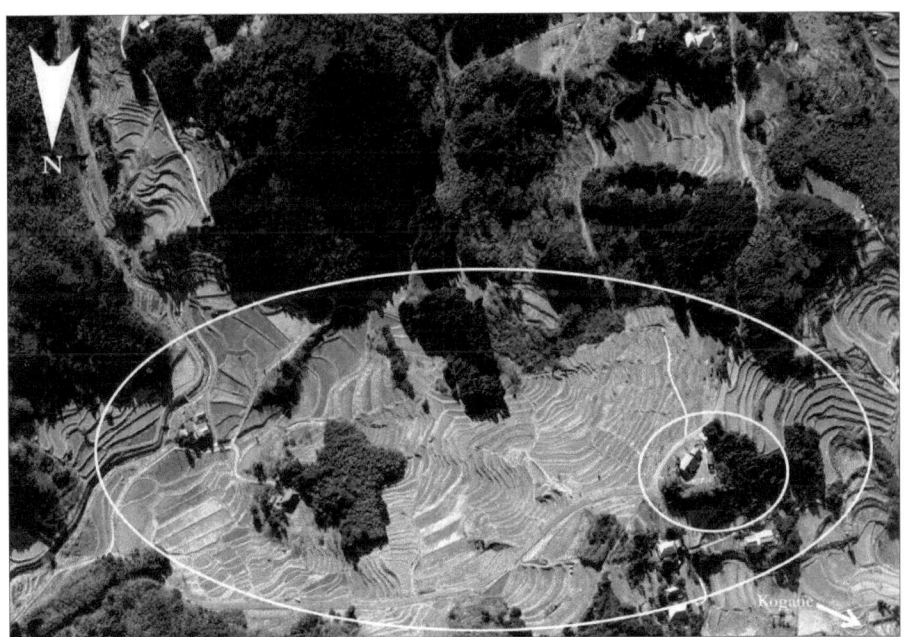

Abb. 14.   Luftbild des Untersuchungsgebietes, 15. Jänner 2004 (Quelle: Keiyō Survey Co).
図 14.   研究区の航空写真、2004年1月15日撮影　(Keiyō Survey Co)。
Fig. 14.   Aerial photo of the study site, taken on 15. January 2004 (source: Keiyō Survey Co).

Eine Autostraße führt am obersten Rand des Reisterrassenkomplexes den Berg entlang hinauf (siehe Abb. 14, unterer Bildrand, mittig). Am höchsten Punkt der Terrassen befinden sich ein Parkplatz und das Besucherzentrum/Vereinshaus, der 'Reisterrassen-Klub' (*tanada kurappu*, siehe Abb. 14, kleine weiße Umrandung). Die nächste Ortschaft Kogane (Gemeinde Hiratsuka) liegt in unmittelbarer Nähe rückwärts des Klubhauses auf der Rückseite des Hügels und besteht aus weniger als 20 Haushalten (KAMOGAWA CITY 2004). Die BewohnerInnen sind großteils PensionistInnen (TB[7]: 2005 – 2006, 11/2010). 1999 wurde Ōyamasenmaida vom Japanischen Ministerium für Landwirtschaft, Forst und Fischerei als eine der hundert schönsten Reisterrassen des Landes ausgezeichnet (AGENCY FOR CULTURAL AFFAIRS 2003B). Bereits 1997, also schon vor dieser Auszeichnung, hatten sich die EigentümerInnen und die lokale Bevölkerung zusammengetan und einen Verein, die NPO 'Ōyamasenmaida Schutzvereinigung', gegründet, da sie erkannt hatten, dass ihre Reisterrassen etwas ganz Besonderes sind, das es zu erhalten gilt (IG: DIREKTOR, 18. JUNI 2005).

In Ōyamasenmaida werden fünf verschiedene Programme für Sojabohnen, Reiswein, Baumwoll-Indigo und Reis angeboten, für die sieben durchstrukturierte kollektive Arbeitstermine (inklusive Erntefest) für eine Saison von Mai bis Oktober anberaumt sind (EI: PÄCHTER A, 2. NOVEMBER 2005). Aufgrund des starken TeilnehmerInnen-Andranges (Tōkyō-Nähe und steigender Bekanntheitsgrad – sogar das japanische Kaiserehepaar war schon da [IMPERIAL HOUSEHOLD AGENCY 2011]) wurde die ursprüngliche Fläche von 3,5 ha (375 Terrassen) auf 4,5 h (415 Terrassen) für das Ownership-System ausgeweitet (ŌYAMASENMAIDA CULTURAL LANDSCAPE PRESERVATION COMMITTEE 2006) – teilweise auch auf ein Areal etwa einen guten Kilometer von Ōyamasenmaida entfernt, bei Ohata (UNVERÖFFENTLICHTE DATEN; TB: 16. OKTOBER 2005).

Die Auswahl von Ōyamasenmaida als Untersuchungsregion erfolgte aufgrund der zuvor dort selbst durchgeführten vegetationsökologischen Forschungen zur Biodiversität in einer traditionellen Reisterrassenlandschaft Japans. Weiteres spielten auch andere forschungspragmatische Gründe, wie die Nähe zu Tōkyō, die Möglichkeit der Erreichbarkeit mit öffentlichen Transportmitteln (das letzte Stück den Berg hinauf musste zu Fuß zurückgelegt werden), bestehende Kontakte sowie vorhandene Studien und Daten, eine Rolle. Die explorative Untersuchung des *Tanada*-Ownership-Systems, die später einsetzte, wurde ebenso dort durchgeführt, da ich inzwischen die Gegend gut kannte und eine Vertrauensbasis zur lokalen Bevölkerung aufgebaut hatte. Aufgrund der attraktiven (bereits national ausgezeichneten) Landschaft eignete sich der Ort außerdem ausgezeichnet für die Untersuchung der Bedeutung von Landschaftsästhetik als Motivation für Kulturlandschaftsschutzaktivitäten. Auch war es mir wichtig, ein gut etabliertes und funktionierendes Kulturlandschaftserhaltungssystem zu studieren, was bei Ōyamasenmaida, das sich bereits im achten Jahr befand, gegeben war.

3. Methodik

Methodische Zugangsweise für die Dissertation war die einer 'case study'. Fallstudien werden v.a. in den Sozialwissenschaften für explorative Forschungen eingesetzt: *„A case study is an empirical inquiry that investigates a contemporary phenomenon in depth and within its real-life context, especially when the boundaries between phenomenon and context are not clearly evident"* (YIN 2003, S. 13).

Mehr als 50 Tage (von Dezember 2004 – Mai 2006) verbrachte ich in Ōyamasenmaida und wohnte dabei Gruppentreffen, Arbeitseinsätzen (z.B. Reispflanzen, Unkrautjäten), traditionellen/kulturellen Festen und Workshops, aber auch dem ganz 'normalen Arbeitsalltag' bei; einerseits direkt im Reisterrassen-Klub, andererseits aber auch bei lokalen AkteurInnen zu Hause.

Die Daten wurden mittels Experten-Interviews 'EI' mit dem Direktor des Ownership-Systems und eines langjährigen Reisfeld-Pächters (EI: PÄCHTER A), Teilnehmender

Beobachtung 'TB', Informeller Gespräche[9] 'IG' sowie einer strukturierten schriftlichen Befragung (mittels Fragebogen, der mit der Post versandt bzw. im Reisterrassenklub aufgelegt wurde) erhoben.

Bei der Befragung handelte es sich bei den PächterInnen 'FP' um eine Vollerhebung (453 PächterInnen, Rücklaufquote 55 %). Die BesucherInnen 'FB' von Ōyamasenmaida hingegen wurden vor Ort angesprochen und gebeten, den Fragebogen auszufüllen (184 BesucherInnen). Auch wurden 30 'Haupt'-AkteurInnen[10] (vor allem Einheimische) und die acht EigentümerInnen[11], die ihre Reisterrassen an die NPO vergaben, mittels Fragebogen interviewt. Die Fragebögen wurden aufbauend auf ersten explorativen ExpertenInterviews, teilnehmender Beobachtung und informellen Gesprächen mit PächterInnen und AkteurInnen der NPO entworfen. Da bei den PächterInnen meist keine E-Mail-Adressen bekannt waren, wurden die Fragebögen bei Veranstaltungen ausgeteilt. Die Personen, die so nicht erreicht werden konnten, bekamen den Fragebogen mit einem frankierten Rückkuvert per Post zugesandt. Der Direktor unterstützte die Studie und ermunterte alle in einem Begleitschreiben auf der ersten Seite des Fragebogens, sich an der Studie zu beteiligen. Die Fragebögen wurden auf Englisch entworfen und dann ins Japanische übersetzt. Sie enthalten offene, halboffene und geschlossene Fragen bezüglich Wahrnehmung und Bedeutung von Kulturlandschaft im Allgemeinen und im Speziellen, Motivation und Einsatz für den Kulturlandschaftsschutz, Bewirtschaftungsweise etc. Nach Rücklauf der Fragebögen wurden die offenen Antworten vom Japanischen ins Englische übersetzt und kodiert. Einige Personen antworteten auch auf Englisch oder Deutsch.

---

[9] Zu den informellen Gesprächen 'IG' zähle ich v.a. ero-epische Gespräche vor Ort (siehe GIRTLER 2001), aber auch mündliche Mitteilungen (persönlich, am Telefon oder via E-Mail).
[10] In Ōyamasenmaida helfen bei Veranstaltungen (z.B. beim Reispflanzen oder Erntefest) viele Leute (v.a. Einheimische) mit. Es gibt aber keine genaue Zahl darüber; 30 Personen jedoch wurden mir als besonders aktiv genannt (KIENINGER ET AL. 2011).
[11] Die Reisterrassen von Ōyamasenmaida gehören insgesamt 11 EigentümerInnen. Aber nur 8 Personen nahmen zum Zeitpunkt der Untersuchung (2006) am Ownership-System teil und verpachteten ihre Reisterrassen (IBID.).

Bei den Experteninterviews und den informellen Gesprächen wurden teilweise zusammenfassende Protokolle unmittelbar nach dem Interview/Gespräch auf Englisch erstellt (Methode siehe KUCKARTZ 2010), oder es fand eine abgekürzte Transkription von Audiofiles statt, bei der relevante Teile des Originaltextes wörtlich transkribiert wurden, während der restliche Inhalt paraphrasiert dargestellt und dann auf Englisch übersetzt wurde (IBID.).

4. Kurzdarstellung der Artikel und ihrer Ergebnisse

Artikel 1: Kieninger, P., Holzner, W., Kriechbaum, M. 2009. Biocultural Diversity and Satoyama. Emotions and the fun-factor in nature conservation – A lesson from Japan. Die Bodenkultur 60 (1): 15–21.

Bei dem ersten Artikel handelt es sich um eine Literaturarbeit, die das japanische Naturverständnis beleuchtet, dem europäischen gegenüberstellt und der Frage nachgeht, inwieweit sich das jeweilige Naturkonzept auf den Umgang mit Natur auswirkt bzw. es für den Schutz von Biodiversität 'genützt' werden kann. Dabei wird auch das japanische Konzept von *satoyama* – der traditionellen Kulturlandschaft Japans (siehe 1.2) – näher unter die Lupe genommen. Der Begriff *satoyama* vereint in sich Natur und Kultur und kann gleichbedeutend für biokulturelle Vielfalt verwendet werden. Anders als die relativ wissenschaftlich anmutenden Termini 'Biokulturelle Vielfalt' und 'Kulturlandschaft' hat sich *satoyama* zu einem populären Begriff der Alltagssprache entwickelt, der Emotionen weckt. In Japan wird der Mensch als ein Teil der Natur verstanden und (damit) der menschliche Einfluss in die Natur als nichts Negatives wahrgenommen. Im Gegenteil: Der formende und gestaltende Prozess des Menschen erfährt höchste Wertschätzung. Aus alten Traditionen heraus wird v.a. ganz bestimmten Tier- und Pflanzenarten sowie besonderen Naturerscheinungen (Herbstfärbung, September-Vollmond, Kirschblüte) und herausragend schönen Landschaften Aufmerksamkeit geschenkt. Diese Sensibilität kann insbesondere gut für die Kulturlandschaftserhaltung eingesetzt werden, indem man die TeilnehmerInnen gezielt über diese emotionalen und ästhetischen Werte anspricht und motiviert. Wie auch bei uns, ist der Großteil der japanischen Landschaft Kulturlandschaft. Während in Mitteleuropa tendenziell eher ein segregativer Naturschutz betrieben wird, ist er in Japan vielmehr integrativ und partizipativ. Natürlich ist eine solche Zugangsweise, z.B. in Naturreservaten, nicht immer durchgehend von Vorteilen. Ein großer Vorteil jedoch ist, dass durch die emotionale Ansprache mehr Personen mobilisiert werden können, sich für ihre eigene (!) Natur und Biodiversität einzusetzen. Denn was man liebt, das schützt man.

Artikel 2: Kieninger, P.R., Yamaji, E., Penker, M. 2011. Urban people as paddy farmers: The Japanese Tanada Ownership System discussed from a European perspective. Renewable Agriculture and Food Systems 26 (4): 328–341.

Im zweiten Artikel untersuchen wir an einer Fallstudie in Ōyamasenmaida, wie Kulturlandschaftserhaltung von und mit Nicht-LandwirtInnen betrieben wird. In Ōyamasenmaida gibt es seit 1997/1998 ein *Tanada*-Ownership-System (siehe 1.5), in dem sich v.a. GroßstädterInnen aus Tōkyō und Chiba beteiligen. In unserer empirischen Studie gehen wir der Frage nach, warum die Einheimischen und LandeigentümerInnen das *Tanada*-Ownership-System gegründet haben, wie es aufgebaut ist und wer sich darin warum und wie engagiert. Die Ergebnisse zeigen, dass das Hauptteilnahmemotiv der städtischen PächterInnen die große Liebe zur Landschaft, speziell zu den Reiseterrassen, ist. Als reines Produktionssystem von Reis, Soja und Reiswein hat das Ownership-System in Ōyamasenmaida keine große Bedeutung, da der Jahresbeitrag verhältnismäßig hoch ist (die Produkte wären billiger im Handel zu erwerben). Die selbst erzeugten Produkte haben aber dennoch einen hohen ideellen Wert für die PächterInnen und sind ein wichtiger Motivator des Programms. Für die LandeigentümerInnen und die lokale Bevölkerung ist das Hauptmotiv, sich aktiv am Ownership-System zu engagieren, der Austausch zwischen Land- und Stadtbevölkerung, von dem sie sich erhoffen, dass – über einen längeren Zeitraum gesehen – wieder Leute aufs Land zurückziehen und dieses beleben. Diese Stadt-Land-Partnerschaften sind auch in staatlichen Regierungsprogrammen eine häufig genannte Strategie, um der Nutzungsaufgabe der traditionellen japanischen Kulturlandschaft (und damit dem Verlust von Biodiversität und biokultureller Vielfalt) zu begegnen.

Artikel 3: Kieninger, P.R., Penker, M., Yamaji, E. 2012. Esthetic and spiritual values motivating collective action for the conservation of traditional rural landscapes – A case study of rice terraces in Japan. Renewable Agriculture and Food Systems DOI: http://dx.doi.org/10.1017/S1742170512000269: 1–16.

Der dritte Artikel baut unmittelbar auf dem zweiten auf und vergleicht die dort bereits betrachtete Untersuchungsgruppe der PächterInnen mit einer weiteren Gruppe, nämlich jener der BesucherInnen. Zentrale Frage war, welche Bedeutung Landschaftsästhetik und spirituelle Werte haben, sich am Ownership-System zu engagieren (PächterInnen) bzw. Ōyamasenmaida zu besuchen (BesucherInnen). Landschaftliche Schönheit spielt bei beiden Untersuchungsgruppen eine wesentliche Rolle. Dennoch zeigt sich, dass die PächterInnen die Landschaft als signifikant schöner wahrnehmen als die BesucherInnen. Auch ist das ökologische Interesse der PächterInnen signifikant größer als das der BesucherInnen. Der Glaube an eine beseelte Natur ('belief in nature spirits in the mountains, valleys, rice terraces, orchards, streams, lakes, plants, and trees') ist bei beiden Gruppen gleichermaßen überraschend hoch und unterscheidet sich nicht wesentlich voneinander. Signifikante Unterschiede zwischen den beiden Untersuchungsgruppen werden bezüglich des Statements sichtbar, dass sie immer daran glauben. Noch eindeutiger werden diese Unterschiede, wenn man nur die Frauen betrachtet: Signifikant mehr Pächterinnen als Besucherinnen glauben immer an die Existenz einer beseelten Natur. Ob das höhere spirituelle Bewusstsein durch die oft vieljährige landwirtschaftliche Tätigkeit und die damit langfristige Auseinandersetzung mit der Landschaft erst entstanden ist oder vielmehr bereits anfänglicher Motivator für das Engagement war, konnten wir mit der angewandten Methode einer Querschnittsuntersuchung nicht endgültig klären. Eine Korrelation der Ausprägungen mit der Teilnahmedauer konnte jedenfalls nicht nachgewiesen werden.

5. Schlussfolgerungen und Ausblick

In meiner Dissertation habe ich mich mit einem ganz konkreten Freiwilligen-Modell zur Erhaltung der Kulturlandschaft und der damit verbundenen Biodiversität in Japan, dem sogenannten *Tanada*-Ownership-System, auseinandergesetzt. In Mitteleuropa basiert die Motivation zur Beteiligung in Naturschutzaktivitäten oft auf naturwissenschaftlichen Erkenntnissen oder ExpertInnenmeinungen, wie z.B. zum Verlust von Biodiversität und Arten. Die Schlussfolgerung dieser Doktorarbeit geht jedoch dahin, dass es erfolgversprechend sein könnte, die Menschen, so wie es in Japan oft gehandhabt wird, auch über emotionale und ästhetische Motive anzusprechen, um möglichst viele Personen mit ins Boot' zu holen. Emotional involvierten Personen könnten in einem zweiten Schritt auch ökologische Zusammenhänge näher gebracht werden. Dass das gelingen kann, zeigt das untersuchte *Tanada*-Ownership-System in Japan.

Das *Tanada*-Ownership-System ist zum Schutz der traditionellen Reislandschaften Japans entstanden. Langfristig erhoffen sich die BesitzerInnen und die lokale Bevölkerung eine Belebung und Wiederbesiedelung des ländlichen Raumes (durch, wenn möglich, junge Leute). Das *Tanada*-Ownership-System soll eine Brücke schlagen und den Wunsch für ein Leben auf dem Land wecken. Ob diese Rechnung aufgeht, ist momentan noch nicht ablesbar. Es zeigen sich jedoch Tendenzen, dass es derzeit v.a. eher ältere Personen sind, die, nachdem sie aus der Erwerbstätigkeit ausgeschieden sind, von der Stadt aufs Land ziehen.

Durch den Non-Profit-Gedanken der Schutzorganisation, der die EigentümerInnen im *Tanada*-Ownership-System angehören, die straff-kollektive Organisation (es gibt kaum Spielraum, eigene Arten und Sorten anzubauen, individuelle Anbaumethoden zu testen oder nach persönlicher Zeiteinteilung zu kommen und zu gehen) und den relativ geringen Ertrag zu einem relativ hohen Preis unterscheidet sich das *Tanada*-Ownership-System von auf die Eigenproduktion ausgerichteten (Civic/Urban) Farming-Projekten wie Gemeinschaftsgärten, Selbsterntefelder oder Schrebergärten. Die 'Rent a Weinstock'-Idee, die in Österreich im Weinviertel und Burgenland umgesetzt wird, ähnelt in ihrem Aufbau dem *Tanada*-Ownership-System (die TeilnehmerInnen können eine bestimmte Anzahl Rebstöcke pachten, sind bei Fixterminen, wie Rebschnitt, Aufbinden, Laubarbeit, Lese, Weinsegnung und Etikettierung, mit dabei und erhalten am Ende der Saison eine gewisse Anzahl an Flaschen Wein), scheint sich aber in den dahinterliegenden Motiven der Beteiligten vom japanischen Reisterrassen-Pachtsystem zu unterscheiden.

Mehr darüber herauszufinden, wäre ein künftiges Forschungsprojekt, das ich gerne in Angriff nehmen würde. Forschungsbedarf und Interesse bestehen auch bezüglich des Zusammenhangs von Spiritualität und der Beteiligung in Naturschutzaktivtäten, um die 'Henne-Ei'-Frage zu klären: Was war zuerst? War die Spiritualität mit ein Grund für das Engagement, oder entwickelte sich die Spiritualität erst durch die Beschäftigung mit und in der Landschaft? Eine Klärung dieser Frage könnte dazu beitragen, Personen künftig für Naturschutz- und Landschaftspflegeaktivitäten gezielter anzusprechen und längerfristig dafür zu motivieren. Dies gilt natürlich auch für andere Motive. Ein Forschungsprojekt, das das Auffinden von Schlüsselfaktoren zur Mobilisierung von AkteurInnen für eine nachhaltige Kulturlandschaftserhaltung und ihrer biokulturellen Vielfalt zum Ziel hat (in Japan und Österreich), ist daher geplant. Als Modellgebiet soll unter anderem auch wieder Ōyamasenmaida dienen.

6. Literatur

Agency for Cultural Affairs, Japan, Department of Cultural Properties, Division of Monuments and Sites (Hg.). 2003a. The Report of the Study on the Protection of Cultural Landscapes Associated with Agriculture, Forestry and Fisheries. http://www.bunka.go.jp/english/pdf/nourinsuisan.pdf. Zuletzt eingesehen am 19.07.2012.

Agency for Cultural Affairs, Department of Cultural Properties, Division of Monuments and Sites, Japan (Hg.). 2003b. Nihon no bunkatekikeikan. Nōrinsuisangyō ni kanren suru bunkatekikeikan no hogo ni kansuru chyōsakenkyū hōkokusho (auf Japanisch).

Bochorodycz, B. 2010. The Changing Patterns of Policy Making in Japan Local Policy Initiative of Okinawa Prefecture in the 1990s. Adam Mickiewicz University, Series Orientalistyka 2, Wydawnictwo Naukowe UAM, Poznań.

Daily Yomiuri Online. 2011. Govts want storks to bring boom / Local authorities hope to boost tourism, agriculture by breeding birds http://www.yomiuri.co.jp/dy/national/T110219003822.htm. February 20, 2011. Zuletzt eingesehen am 31.07.2012.

FAO. 2008. Conservation and Adaptive Management of Globally Important Agricultural Heritage Systems (GIAHS), Terminal Report, Project Symbol: UNTS/GLO/002/GEF, Project ID: 137561. FAO, Rome.

Fujioka, M., Lee, S.D., Kurechi, M., Yoshida, H. 2010. Bird use of paddy fields in Korea and Japan. Waterbirds 33: 8–29.

Fujioka, M., Yoshida, H. 2001. The Potential and Problems of Agricultural Ecosystems for Birds in Japan. Global Environmental Research 5 (2): 151–161.

Geographical Survey Institute 2000. Changes in wetland area, Japan. http://www1.gsi.go.jp/geowww/lake/shicchimenseki2.html. Zuletzt eingesehen am 25.07.2012 (auf Japanisch).

Girtler, R. 2001. Methoden der Feldforschung. 4. Auflage. Böhlau, Wien-Köln-Weimar.

Hayashi, A. 2002. Finding the Voice of Japanese Wilderness. International Journal of Wilderness 8 (2): 34–37.

Holzner, W. 1983. Man's Impact on vegetation in Japan and Central Europe. In: W. Holzner, M.J.A. Werger, I. Ikushima (Hg.). Man's impact on vegetation. Dr. W. Junk Publishers, The Hague-Boston-London, S. 341–357.

Imperial Household Agency. 2011. http://www.kunaicho.go.jp/activity/gonittei/01/h22/gonittei-1-2010-3.html. Zuletzt eingesehen am 19.07.2012 (auf Japanisch).

Iwabuchi, S., Kurechi, M., Kashiwagi, M. 2010. Biodiversity in Rice Paddies.

Iwata, Y., Fukamachi, K., Morimoto, Y. 2010. Public perception of the cultural value of Satoyama landscape types in Japan. Landscape Ecology Engineering 7 (2): 173–184.

Ishii, M., Nakamura, Y. 2012. Development and Future of Insect Conservation in Japan. Chapter 15. In: T.R. New (Hg.). Insect Conservation: Past, Present and Prospects. Springer Science+Business Media Dordrecht, DOI 10.1007/978-94-007-2963-6_15, 339–357.

Japanese Association for Wild Geese Protection. 2005. Fuyumizutanbo, environmentally sound rice field. Winter-flooded Rice Fields. Leaflet, Kuihara City.

Japan Times. 2011. Editorial: Creating special economic zones (November 30, 2011). http://www.japantimes.co.jp/text/ed20111130a2.html. Zuletzt eingesehen am 19.07.2012.

Kalland, A. 1995. Culture in Japanese Nature. In: O., Bruun, A., Kalland (Hg.). Asian perceptions of nature. A critical approach. Nordic Institute of Asian Studies, Studies in Asian Topics, No. 18. Curzon Press, Richmond (UK), S. 243–257.

Kamogawa City. 2004. Kamogawashi tōkeishyo – heisei 16 nenban (auf Japanisch).

Kato, M. 2001. 'Satoyama' and Biodiversity Conservation: 'Satoyama' as important Insect Habitats. Global Environ. Research 5 (2): 135–149.

Kellert, S.R. 1995. Concepts of Nature East and West. In: M. Soulé, G. Lease (Hg.). Reinventing Nature? Responses to Postmodern Deconstruction. Island Press, San Francisco, S. 103–121.

Kellert, S.R. 1993. Attitudes, Knowledge, and Behavior Toward Wildlife Among the Industrial Superpowers: United States, Japan, and Germany. Journal of Social Issues 49 (1): 53–69.

Kellert, S.R. 1991. Japanese Perception of Wildlife. Conservation Biology 5 (3): 297–308.

Kieninger, P.R., Penker, M., Yamaji, E. 2012. Esthetic and spiritual values motivating collective action for the conservation of traditional rural landscapes – A case study of rice terraces in Japan. Renewable Agriculture and Food Systems. Renewable Agriculture and Food Systems DOI: http://dx.doi.org/10.1017/S1742170512000269: 1–16.

Kieninger, P.R., Yamaji, E., Penker, M. 2011. Urban people as paddy farmers: The Japanese Tanada Ownership System discussed from a European perspective. RAFS 26 (4): 328–341.

Kitazawa, T., Ohsawa, M. 2002. Patterns of species diversity in rural herbaceous communities under different management regimes, Chiba, central Japan. Biological Conservation 104: 239–249.

Kobori, H. 2009. Current trends in conservation education in Japan. Biological Conservation 142: 1950–1957.

Kobori, H., Primack, R.B. 2003. Participatory Conservation Approaches for Satoyama, the Traditional Forest and Agricultural Landscape of Japan. Ambio 32 (4): 307–311.

Kohsaka, R., Flintner, M. 2004. Exploring forest aesthetics using forestry photo contests: case studies examining Japanese and German public preferences. Forest Policy and Economics 6: 289–299.

Konami, H. 2003. A Framework for Urban Revitalization and Sustainable Development. http://konamike.net/hiro/papers/230212surp.htm. Zuletzt eingesehen am 19.07.2012.

Konishi, M. 2004. Japan, Land der Libellen. Nipponia 29: 25.

Koji, S., Nakamura, A., Nakamura, K. 2012. Demography of the Heike firefly *Luciola lateralis* (Coleoptera: Lampyridae), a representative species of Japan's traditional agricultural landscape. Journal of Insect Conservation 1–9.

Kuckartz, U. 2010. Einführung in die computergestützte Analyse qualitativer Daten. VS Verlag für Sozialwissenschaften, Wiesbaden.

MAFF – Japanese Ministry of Agriculture, Forestry and Fisheries. 2011 (Hg.). FY2010 Annual Report on Food, Agriculture and Rural Areas in Japan. Summary. http://www.maff.go.jp/e/annual_report/2010/pdf/e_all.pdf. Zuletzt eingesehen am 20.07.2012.

MAFF – Japanese Ministry of Agriculture, Forestry and Fisheries. 2008a. Annual Report on Food, Agriculture and Rural Areas in Japan FY 2008. Policies on Food, Agriculture and Rural Areas in Japan FY2007. Summary (Provisional Translation). http://www.maff.go.jp/e/annual_report/2008/pdf/e_all.pdf. Zuletzt eingesehen am 20.07.2012.

MAFF – Japanese Ministry of Agriculture, Forestry and Fisheries. 2008b. Annual Report on Food, Agriculture and Rural Areas in Japan FY 2007. http://www.maff.go.jp/e/annual_report/2007/pdf/e_all.pdf. Zuletzt eingesehen am 20.07.2012.

Maruyama, Y. 2006. Pluralism and Universality of Environmental Discourse: The Dilemma Between Damage Caused by the Japanese Macaque and its Protection. International Journal of Japanese Society 15: 55–68.

MIC – Japanese Ministry of Internal Affairs and Communications, Statistics Bureau, Director-General for Policy Planning (Statistical Standards) & Statistical Research and Training Institute. 2010. Statistical Handbook of Japan 2010. Chapter 2. http://www.stat.go.jp/english/data/handbook/c02cont.htm#cha2_1. Zuletzt eingesehen am 20.07.2012.

Mie, A. 2012. Japanese Crested Ibis. Efforts to save Japanese crested ibis take flight. Japantimes. May 8, 2012. http://www.japantimes.co.jp/text/nn20120508i1.html. Zuletzt eingesehen am 24.03.2013.

Miyaura, T. 2009. Satoyama – A place for preservation of biocultural diversity and environmental education. Die Bodenkultur – Journal for Land Management, Food and Environment 60 (1): 23–29.

MOE – Japanese Ministry of the Environment, Nature Conservation Bureau (Hg.). 2010a. Biodiversity is Life. Biodiversity is our Life. The National Biodiversity Strategy of Japan 2010.

MOE – Japanese Ministry of the Environment, Nature Conservation Bureau, Biodiversity Center of Japan (Hg.). 2010b. Biodiversity of Japan. A Harmonious Coexistence between Nature and Humankind. Heibonsha Ltd., Tōkyō.

MOE – Japanese Ministry of the Environment. 2010c. Payments for ecosystem services (PES) – An Introduction of good practices in Japan. Kabukuri-numa and sourrounding rice paddies. http://www.biodic.go.jp/biodiversity/shiraberu/policy/pes/en/satotisatoyama/satotisatoyama01.html. Zuletzt eingesehen am 30.07.2012.

MOE – Japanese Ministry of the Environment. 2010d. Payments for ecosystem services (PES) – An Introduction of good practices in Japan. Restoring rice paddy habitats to reintroduce the White stork in Toyooka City. http://www.biodic.go.jp/biodiversity/shiraberu/policy/pes/en/satotisatoyama/satotisatoyama02.html. Zuletzt eingesehen am 27.07.2012.

MOE – Japanese Ministry of the Environment. 2010e. Payments for ecosystem services (PES) – An Introduction of good practices in Japan. Reintroducing the crested ibis and rice production. http://www.biodic.go.jp/biodiversity/shiraberu/policy/pes/en/satotisatoyama/satotisatoyama03.html. Zuletzt eingesehen am 27.07.2012.

Morimoto, Y. 2011. What is Satoyama? Points for discussion on its future direction. Landscape and Ecological Engineering 7 (2): 163–171.

Murota, Y. 1985. Culture and Environment in Japan. Environmental Management 9 (2): 105–112.

Nagasawa, K. 2008. The Mountains, Rivers, Grasses and Trees Will All Become Buddhas – The Japanese View of Nature And Religion. In: Biodiversity Network Japan (Hg.). Conserving Nature. A Japanese perspective. S. 32–35.

National Federation of Land Improvement Association. 2012. http://www.inakajin.or.jp/kikin/tanada/tanada_list.html#bunrui. Zuletzt eingesehen am 19.07.2012.

Natori, Y., Fukui, W., Hikasa, M. 2005. Empowering nature conservation in Japanese rural areas: a planning strategy integrating visual and biological landscape perspectives. Landscape and Urban Planning 70:315–324.

Natuhara, Y. 2012 Ecosystem services by paddy fields as substitutes of natural wetlands in Japan. Ecological Engineering. http://dx.doi.org/10.1016/j.ecoleng.2012.04.026,

Numata, M. (Hg.) 1974. The flora and vegetation of Japan. Kodansha, Tōkyō & Elsevier, Amsterdam-London-New York.

Ohsawa, M., Kitazawa, T. 2009. Biocultural diversity and functional integrity of Japan´s rural landscape. Die Bodenkultur – Journal for Land Management, Food and Environment 60 (1): 31–40.

Oyadomari, M. 1989. The Rise and Fall of the Nature Conservation Movement in Japan in Relation to Some Cultural Values. Environmental Values 13 (1): 23–33.

Ōyamasenmaida Cultural Landscape Preservation Committee. 2006. Ōyama no senmaida bunkatekikeikan hozon katsuyō keikaku. Kabushiki kaisha koa, Kamogawashi (auf Japanisch).

Primack, R., Kobori, H., Mori, S. 2000. Dragonfly Pond Restoration Promotes Conservation Awareness in Japan. Conservation Biology 14: 1553–1554.

Saito, Y. 2010. Future Directions for Environmental Aesthetics. Environmental Values 19: 373–391.

Saito, Y. 2002. Scenic National Landscapes: Common Themes in Japan and the United States. Essays in Philosophy 3 (1): Art. 5.

Salvini, P., Teti, G., Spadoni, E., Frediani, E., Boccalatte, S., Nocco, L., Mazzolai, B., Laschi, C., Comandé, G., Rossi, E., Carrozza, P., Dario, P. 2010. An Investigation on Legal Regulations for Robot Deployment in Urban Areas: A Focus on Italian Law. Advanced Robotics 24: 1901–1917.

Satoyama Initiative. s.a. http://satoyama-initiative.org/en/. Zuletzt eingesehen am 22.07.2012.

Sherkat, D.E., Ellison, C.G. 2007. Structuring the Religion-Environment Connection: Identifying Religious Influences on Environmental Concern and Activism. J. for the Scientific Study of Religion 46 (1): 71–85.

Sood, J., Nasu, Y. 1995. Religiosity and nationality: An exploratory study of their effect on consumer behaviour in Japan and the United States. Journal of Business Research, 34: 1–9.

Sprague, D.S., Iwasaki, N. 2006. Coexistence and exclusion between humans and monkeys in Japan: Is either really possible? Ecological and Environmental Anthropology 2 (2): 30–43.

Takeda, M., Amano, T., Katoh, K., Higuchi, H. 2006. The habitat requirement of the Genji-firefly *Luciola cruciata* (Coleoptera:Lampyridae), a representative endemic species of Japanese rural areas. Biodiversity and Conservation 15: 191–203.

Takeuchi, K. 2010. Rebuilding the relationship between people and nature: the Satoyama Initiative. Ecological Research 25: 891–897.

Takeuchi, K. 2003. The Nature of Satoyama Landscapes. In: K. Takeuchi, R.D. Brown, I. Washitani, A. Tsunekawa, M. Yokohari (Hg.). Satoyama – The Traditional Rural Landscape of Japan. Springer-Verlag, Tōkyō, S. 9–16.

Takeuchi, K. 2001. Nature Conservation Strategies for the 'SATOYAMA' and 'SATOCHI', Habitats for Secondary Nature in Japan. Global Environmental Research 5 (2): 193–198.

Washitani, I. 2007. Restoration of Biologically-diverse Floodplain Wetlands Including Paddy Fields. Global Environ. Research 11: 135–140.

Washitani, I. 2001. Traditional Sustainable Ecosystem 'SATOYAMA' and Biodiversity Crisis in Japan : Conservation Ecological Perspective. Global Environmental Research 5 (2): 119–133.

White, L. 1967. The Historic Roots of Our Ecologic Crisis. Science 183: 1203–1207.

Woodrum, E., Wolkomir. M.J. 1997. Religious Effects on Environmentalism. Sociological Spectrum 17: 223–234.

Yamada, S., Okubo, S., Kitagawa, Y., Takeuchi, K. 2007. Restoration of weed communities in abandoned rice paddy fields in the Tama Hills, central Japan. Agriculture, Ecosystems and Environment 119: 88–102.

Yamaji, E. 2006. Enjoyment of rural amenities by ownership program of rice terraces. Rural Planning Association 25: 206–212 (auf Japanisch).

Yin, R.K. 2003. Case study research: design and methods. 3rd ed. Applied social research methods series, Volume 5. Sage Publications, Thousand Oaks, London, New Delhi.

Yugami, K. 2008. Job Creation by Local Initiatives: Effects of Special Zones for Structural Reform. Japan Labor Review 5 (1): 101–121.

7. Dank

Von ganzem Herzen möchte ich allen Menschen danken, die zur Entstehung dieser Doktorarbeit beigetragen haben …

… an erster Stelle seien hier meine drei BetreuerInnen zu nennen. Meine Dissertation nahm bei Prof. **Wolfgang Holzner** (Institut für Integrative Naturschutzforschung, BOKU) ihren Anfang und ich erinnere mich noch gut, wie er anfänglich versuchte, mir das 'Projekt Japan' auszureden, mich dann aber, als er an meinem Dickkopf scheiterte, nach allen Kräften unterstützte. Der Austausch mit ihm, einem Japan-Kenner, war für mich nicht nur interessant, sondern auch essenziell für die Arbeit. Lieber Wolfgang, danke Dir für alles! Prof. **Marianne Penker** (Institut für Nachhaltige Wirtschaftsentwicklung, BOKU) bin ich auch zu allergrößtem Dank verpflichtet. Dafür, dass sie, nachdem sich meine Doktorarbeit im Laufe des zweijährigen Forschungsaufenthaltes in Japan thematisch gedreht hatte, ohne Zögern die Mit- und später die Hauptbetreuung übernommen hat, sowie für all ihre Ratschläge, Hinweise, Korrekturen, ihre Aufmunterung und den Glauben in mich und die Arbeit. Tausend Dank Dir Marianne! Prof. **Eiji Yamaji** (Division of Environmental Studies, Department of International Studies, Tōkyō University) hat mir in Japan bei den empirischen Erhebungen zur Seite gestanden, auf denen meine Doktorarbeit fußt. Thank you very much Yamaji-Sensei!

… allen Mitgliedern und FreundInnen der **NPO 'Ōyamasenmaida Schutzvereinigung'**, insbesondere **Mitsuji Ishida, Yoshiko** und **Etsuko Sudo, Hitomi Taira, Mikiko Matumoto, Junko Nagamura, Masako** und **Hisao Kawana, Kakichi Omura, Chouji Suzuki** und **Michiko Yamane** (siehe S. 27–29), für die Unterstützung, die sie mir, einem *'gaijin'*, sowohl während als auch nach Ende meiner Feldarbeiten entgegenbrachten. Im Klubhaus in Ōyamasenmaida gab es immer ein offenes Ohr für mich und meine Anliegen, einen Gratis-Schlafplatz und Verpflegung. どもありがとうございました！

… ferner allen **ReisfeldpächterInnen** und **BesucherInnen** Ōyamasenmaidas für ihre Beteiligung an meiner Fragebogenaktion.

… Dipl.-Ing. Dr. **Kentaro Aoki** (UNIDO/IIASA, Wien), PhD **Nobuko Morishita** (Institut für Orientalische Schriften, Russische Akademie der Wissenschaften), Mag. Dr. **Isabelle Prochaska** (Japanologie, Institut für Ostasienwissenschaften, Universität Wien), PhD **Nobuhiko Sawai** (Gunma Universität, Hochschule für Medizin), PhD **Ayako Toko** (WWF Japan) und PhD **Yuko Yoneda** (Präfektur Universität Kyoto), die mich bei der Übersetzung der Interviews und der offenen Antworten der Fragebögen vom Japanischen ins Deutsche bzw. Englische unterstützt haben.

… meiner Kollegin Mag. Dr. **Katharina Bardy** (Institut für Integrative Naturschutzforschung, BOKU) für ihre Hilfe bei der statistischen Auswertung der Daten sowie dem gesamten Institut für das Interesse an meiner nicht nur geographisch weit entfernt liegenden Forschungsarbeit. Nennen möchte ich hier auch meine ehemaligen KollegInnen von der Tōkyō University (Laboratory of Biosphere Function, Institute of Environmental Studies, Graduate School of Frontier Sciences) für die Unterstützung bei den Feldarbeiten, v.a. PhD **Nobuki Kawano**, PhD **Tetsuya Sano** und PhD **Pema Wangda** sowie Prof. **Masahiko Ohsawa** für seine Hilfe bei der Auswahl des Untersuchungsgebietes.

… dem Japanischen Ministerium für Bildung, Kultur, Sport, Wissenschaft und Technologie (**MEXT**) und dem Deutschen Akademischen Austauschdienst (**DAAD**) für die Finanzierung meiner Forschungsarbeiten in Japan.

… Mag. Dr. **Birgit Staemmler** (Philosophische Fakultät, Japanologie, Eberhard Karls Universität Tübingen) für ihre Ratschläge zu Religion und Spiritualität in Japan.

… MSc **Naoki Amako** (Vizedirektor Abteilung Biodiversität, Unterabteilung Naturschutz, Umweltministerium Japan) für seine Hilfe bei der Suche biodiversitätsbezogener Daten in Japan.

… BSc (Hons) **Vivien Landauer und** LL.M. **Marie-Alice Hofmaier** für die Zeit, die sie sich für die Englisch-Korrekturen der Artikel genommen haben.

… den anonymen **RedakteurInnen** der Zeitschriften '**RAFS**' und '**Bodenkultur**' für ihre wertvollen Verbesserungshinweise in den jeweiligen Artikeln sowie Prof. **Erwin Frohmann** (BOKU) und Prof. **Sepp Linhart** (Universität Wien/Japanologie) für ihre **Gutachten**.

… der **Cambridge University Press** und dem **facultas.wuv Universitätsverlag** für die freundliche Genehmigung des Abdrucks der Artikel.

… meiner **Familie** (insbesondere meinen Eltern) und **FreundInnen** für die Durchhalteparolen, als die Dissertation kein Ende nahm.

Danke!

大山千枚田の皆様、どうもありがとうございました！

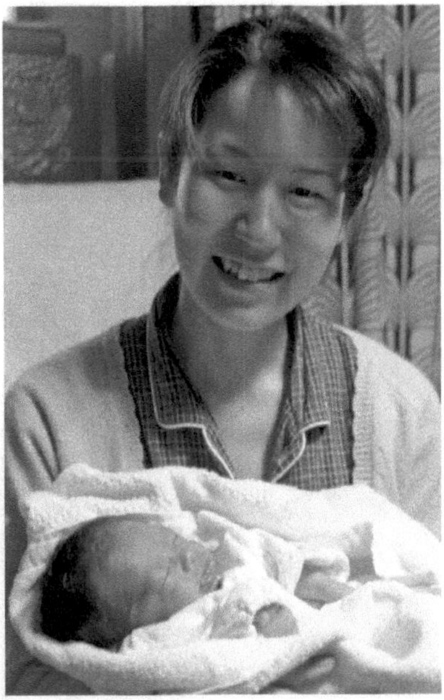

Rahmenschrift Dissertation Pia R. Kieninger

8. Wissenschaftliche Publikationen

Artikel 1: Kieninger, P., Holzner, W., Kriechbaum, M. 2009. Biocultural Diversity and Satoyama. Emotions and the fun-factor in nature conservation – A lesson from Japan. Die Bodenkultur 60 (1): 15–21.

Artikel 2: Kieninger, P.R., Yamaji, E., Penker, M. 2011. Urban people as paddy farmers: The Japanese Tanada Ownership System discussed from a European perspective. Renewable Agriculture and Food Systems 26 (4): 328–341.

Artikel 3: Kieninger, P.R., Penker, M., Yamaji, E. 2012. Esthetic and spiritual values motivating collective action for the conservation of traditional rural landscapes – A case study of rice terraces in Japan. Renewable Agriculture and Food Systems DOI: http://dx.doi.org/10.1017/S1742170512000269: 1–16.

Rahmenschrift Dissertation Pia R. Kieninger

Artikel 1: Kieninger, P., Holzner, W., Kriechbaum, M. 2009. Biocultural Diversity and Satoyama. Emotions and the fun-factor in nature conservation – A lesson from Japan. Die Bodenkultur 60 (1): 15–21.

Biocultural Diversity and Satoyama. Emotions and the fun-factor in nature conservation – A lesson from Japan

P. Kieninger, W. Holzner and M. Kriechbaum

Biokulturelle Diversität und Satoyama. Emotionen und Spaßfaktor im Naturschutz – Beispiele aus Japan

1 Introduction

"Biodiversity virtually has become a cult concept" (COUNCIL OF EUROPE, 1996), and is a common topic of conversation among people concerned with environmental topics. Many conservation policies, programs and projects, laws and regulations have been designed for the conservation of biodiversity, but not much success can be reported; publications are still dominated by reports on the continual losses of biodiversity worldwide. And, if one considers that the diversity of the living world in space and time is a meta-concept with practically indefinite attributes (NORTON, 1994), it is questionable whether science can be helpful here at all. The bio-scientific approach to conservation issues will have to be amended or supported by economic and social concerns, which consider attitudes, emotions and values as important factors in the system.

The conclusion of an international group of experts discussing the possibilities of achieving the goals of the Convention on Biological Diversity (Rio 92) was: "*Successful implementation of the CBD depends on a complex interplay of ecological processes, culture, economic and social concerns.* Be-

Zusammenfassung

Die Erhaltung der Biodiversität ist abhängig von komplexen Wechselwirkungen zwischen ökologischen, ökonomischen und kulturellen Prozessen. In diesem Beitrag werden europäische und japanische Konzepte zur Erhaltung der Biodiversität hinterfragt und verglichen. Ein rein wissenschaftlicher Ansatz, wie er in Europa vorherrscht, ist dabei deswegen kontraproduktiv, weil er die Menschen aus der Natur ausschließt und daher verhindert, dass sie für „ihre" Biodiversität Verantwortung übernehmen. Der Ansatz, der in Japan verfolgt wird, gründet in elnem traditionellen Verständnis einer Natur, die sehr stark von der Kultur geprägt ist und zeichnet sich durch integrative und partizipative Strategien aus. Der japanische Begriff Satoyama, der die Kulturlandschaft umschreibt, die Einheit von Natur und Kultur, die daraus entstehende Biodiversität und ihre Bedeutung für unsere Lebensqualität, ist praktisch gleichbedeutend mit dem weltweit aktuellen Konzept der „biokulturellen Diversität", das die Verbindung zwischen der Vielfalt von Natur und Kultur unterstreicht.

Schlagworte: Satoyama, Biodiversität, Biokulturelle Vielfalt, Naturschutz, Kulturlandschaft, Japan.

Summary

Biodiversity depends on a complex interplay of ecological processes and of economy and culture. Public consensus is necessary for sustainable preservation programs. To achieve such programs, conservation concepts that keep humans out of nature and take a merely scientific approach are counterproductive. They discourage citizens from taking responsibility for a nature, which they do not understand and where they are unwanted. The Japanese however, traditionally do not differentiate between untouched nature and nature shaped by culture. The Japanese term, satoyama, stands not only for the traditional rural landscape but also includes biodiversity, social traditions, and emotions.

Keywords: Satoyama, biodiversity, biocultural diversity, nature conservation, cultural landscape, Japan.

cause of this complex dependency the conservation of biodiversity has to be mainstreamed." However, in Europe we seem to be far away from this goal. "Nature" is considered as something different and separate from human affairs, something that is always outside and somewhere else and consequently nature conservation means to keep the people away as much as possible.

In this paper we will firstly discuss some of the reasons for this development and the resulting problems for the preservation of biodiversity. Then we will focus on Japan. We will examine whether the different, rather intimate relations of the Japanese with nature could serve as a clue to explain why some of the nature conservation activities there are different and perhaps more successful than European ones generally are.

How far can examples from Japan be used here as a model for Europe?

2 Biodiversity: a field of science or a political program?

"Biological diversity" is a rather young field of research in which biologists try to acquire a grasp of the manifold variety of natural phenomena. When it was launched as a political slogan in 1986[1] to raise awareness and concern about the losses of plant and animal species among politicians and in the public, "biological diversity" was shortened to "biodiversity": "*It was easy to do: all you do is to take the 'logical' out of 'biological'.*" (ROSEN in TAKACS, 1996). "Biodiversity" is not only a shorter and catchier word. The omission of logics has a deeper meaning: "biodiversity" stands for a phenomenon, that cannot be measured by splitting it up into elements which can be counted, and for the immeasurable human values, feelings and emotions connected with the amazing variety of nature. It is an expression of the longings and desires of our civilization for a kind of paradise lost [2]. "*Biodiversity: We search for ways to preserve it, which means preserving intact rain forests as well as preserving our value systems, our awe and wonder we want all to pass on to future generations, and biodiversity as a term encompasses all of it.*" (TAKACS, 1996).

Besides that, within a conservation context "biodiversity" is a construct of society. "Diversity" (of any kind) cannot be treated as value-neutral, as it is inherently subjective (ESER, 2003). In this political context, "biodiversity" is not an objective concept of science, but a product of the values of certain people in a certain region at a certain time and its decrease is a problem only in so far as it is seen as a problem; it is a problem for people not for nature. Scientific concepts for defining, measuring and evaluating biodiversity for the purpose of biodiversity policy have been elaborated and discussed extensively (e.g. WEIMANN et al., 2003; HOFFMANN et al., 2005). However, this ambivalent concept, balanced between "*science and society, between facts and values*", affords an opportunity to redefine the legitimate boundaries of science in ways that include evaluative statements and political practices (ESER, 2001).

3 A dominant science imposes problems on conservation projects

Biological diversity is a function of space and time and is acting on several different levels, as we see it, from macro-cosmos to the micro-cosmos and indeed at each level there are myriads of problems for science (HEYWOOD, 1994). This means that conservation projects based on science encounter at least some of those "myriads of problems". And these science-made problems often turn out to be major obstacles to the success of a conservation program or project.

Thus, as far as what concerns biodiversity and its preservation, scientists are split into two blocks: many consider biodiversity as a well defined concept and the methods that would be necessary to preserve it as a scientific commonplace. On the other hand, however, there are some critical publications, which suggest that these ideas are naïve: Already in 1992 a team of biologists, after screening the rich amount of biological information available in Great Britain for its suitability to develop conservation strategies, summarized: "*Saving species we will have to rely on luck and intuition; we cannot wait until all data are in, because if we wait there will be nothing left worth conserving.*" (LAWTON et al., 1994).

In the societies of Europe the competence and responsibility for biodiversity preservation issues is in the hands of experts. They do not only suggest or decide what has to be done, they first of all determine which aspect or part of biodiversity is more important than others. The general public has to accept the priorities of the scientists, because they are based on "objective" scientific evidence. If such an approach were to be chosen in a third world development project, it would be heavily criticized as non-participatory and therefore unsustainable. It is necessary to convince the public that their (!) biodiversity is at stake and not the biodiversity of the experts. Preservation must be made into an issue of everybody.

4 Satoyama – linking biological and cultural diversity

The concept of "biocultural diversity" emerged from the relatively recent insight, that the "*two great realms of living diversity are cultural and biological*" (HARMON, 2002) and both are strongly interdependent. There is a growing awareness of the linkage between biological and cultural diversity and the crucial role of both for a sustainable development of human societies and their well-being worldwide.

In this context it is interesting to look at Japanese approaches to the task of biodiversity preservation. A keyword within this context is "*satoyama*", a word which can be understood as representing the unity of nature and culture in rural landscapes. As we have already pointed out, the term "biodiversity" has a strong scientific flavour and the same is the case for "biocultural diversity", even if this happens to be the result of a wrong or incomplete understanding. The term "*satoyama*" has the following advantages:
- It is neutral – the prefix "bio" has been too extensively used in politics and advertising for products from bananas to cosmetics and detergents;
- It is simple, meaning just village and mountain (MIYAURA, this volume);
- It also encompasses feelings and emotions, similar to those conveyed by the Western expressions "homeland", "native country", "land of one's ancestors", "the old country" – for many people a kind of paradise lost or one that should or could be regained.

5 Satoyama – maintaining cultural heritage and biodiversity in Japan

The term "*satoyama*" is very popular in Japan and understood by everyone. *Satoyama* stands for the traditional agricultural landscape of Japan, composed of human settlements at the base of hills, kitchen gardens in front of the houses and bamboo groves at their backside, crop- and paddy fields, coppiced forests and grassland on the slopes, as well as streams, ponds, water reservoirs, temples, shrines and graveyards (MIYAURA, this volume; OHSAWA and KITAZAWA, this volume).

Rice cultivation in Japan is closely intertwined with the traditional rural landscape. In addition, through its thousands of years of tradition, rice cultivation is an immutable part in Japanese life, religion and culture. The Japanese emperor, for example, still celebrates several Shinto ceremonies every year, in order to bless and protect the rice crop. Paddies are also important for the hydrologic balance of the landscape and for erosion control. Furthermore, they act as very important sites for recreation. People from large towns in particular come to the countryside in order to relax from the stressful life in the huge and crowded cities.

Today it is recognized that paddy landscapes play an important role in the preservation of the biodiversity of Japan. They represent, together with streams, ponds, reservoirs, rivers and irrigation ditches the largest area of wetlands, comprising half of all freshwater wetlands (KOBORI and PRIMACK, 2003). Plenty of aquatic species can be found in this environment, as for example most of the frog species of Japan, which use paddy fields as their habitat for pairing, egg deposition, maturing of the eggs, larval growth and adult feeding. But also many other animals depend on paddy fields and other habitats are connected with them. For instance fireflies are a very prominent example. They crawl out of the paddy mud in May. In June, by using their unique luminescence in order to find a partner, they decorate the surroundings of the paddy fields. In July, after depositing eggs on the borders of the rice fields, the firefly's lifetime is running out. The new firefly generation, after hatching from the eggs, overwinters in the mud on the ground of the paddies until May, when a new cycle starts again.

In addition, many birds relay on the existence of the paddy fields and their surroundings, for example, cranes (*Grus* spp.), swans (*Cygnus* spp.), geese (*Anser* spp.), dabbling ducks (*Anas* ssp.), the gray-faced-buzzard-eagle (*Butastur indicus*) and others (FUJIOKA and YOSHIDA, 2001). The white-fronted goose (*Anser albifrons frontalis*) also needs large watersides and rice fields for resting and feeding (KURECHI and IWABUCHI, 2005). It is a winter migratory bird, which breeds in Russia's summer tundra areas. It has been protected by law in Japan since 1971. Recently, to support this and other winter migratory birds, the traditional agricultural method of winter-flooding the rice fields, which was already applied in the Edo (1603–1868) period (KURECHI and IWABUCHI, 2005), has spread out nationwide (KURITA et al., 2006). In this method, water is kept in the paddies during the winter season as well, thus providing a habitat all the year around.

The slopes between paddy fields and forest are another important habitat type, where many rare plants can be found (OHSAWA and KITAZAWA, 2008). These elements of the rural landscape are called "mowing place", *kariage-ba*, because they have been cut regularly in former times. Recently, their management has been nearly given up. Their ecological function however, as that of the whole paddy ecosystem, depends on regular cultivation.

The Japanese government is aware of this. The Japanese Ministry of Agriculture, Forestry and Fisheries (MAFF) implemented the "Direct Payment System" for farmers of hilly and mountainous areas in 2000. The purpose of this system is to overcome the human depopulation and land abandonment trends and to preserve the multi-functionality of *satoyama* (WATANABE, 2003; MINISTRY OF AGRICULTURE, FORESTRY AND FISHERIES, JAPAN, 2005). Hilly mountainous areas of Japan are regions where small-sized terraced rice field farming is predominant. The direct payment system is the first policy in Japan in which subsidies are decoupled from production. Subventions are linked to the steepness of the cultivation area. Terraced paddy fields *(tanada)* with an inclination of 1:20 are subsidized with around 210,000 yen/ha/year (around 1350 €) and terraced paddy fields with an inclination of 1:100 with circa 80,000 yen/ha (around 515 €) (ICHINOSE, 2007).

6 The Japanese concept of "nature"

The Japanese understanding of "nature" is different to that of Europeans. Consequently programs and measures aiming to protect nature show a special Japanese touch (KIENINGER and HOLZNER, in press). Besides that, Japanese people have a particular fondness for "details", such as particular plants, animals or other individual natural objects, as well as for events taking place in nature, like moon-watching, sunrise meditation (a "serious" Mt. Fuji climber will try to reach the top by sunrise!), autumn foliage contemplation or the admiration of the cherry blossoms. This love and veneration managed to survive from ancient times until today. Particularly insects, such as the fireflies *(hotaru)* mentioned previously, especially the *Genji*-firefly – *Luciola cruciata* (TAKEDA et al., 2006), dragonflies (in particular the red dragonfly called *akatonbo*), cicadas, crickets, praying mantis (KONISHI, 2004), as well as *Oryzias latipes*, the Japanese killifish (KOBORI and PRIMACK, 2003), enjoy great popularity among Japanese people. All of them are typical animals of *satoyama*, and thus emphasize the fact that the phenomena of nature which are most esteemed by people, thrive in a rural landscape with very few relics of the original wilderness.

In such a context, the focus of nature conservation on species of high public interest is not a limitation to nature protection. Such species must rather be considered and esteemed as an important stimulus for nature conservation: *"Raising public interest in nature through conserving species of high social interest is crucial in achieving effective conservation of biodiversity."* (TAKEDA et al., 2006).

Conservation awareness in Japan is promoted largely through the esteem for the scenic beauty of certain landscapes. The Japanese sense of landscape beauty is moulded by images of the rural landscape, not by untouched wilderness. Since ancient times, such beautiful places have been renowned as sightseeing spots. One of the most famous sites is the rice terraces *(tanada)* landscape on the steep slopes of Mt. Obasute in Nagano Prefecture. Since the Edo (1603–1868) period this area has been widely know as a moon-watching-point. Many famous woodblock artists, painters and poets came there to create their works. Therefore, this site of the landscape is also very popular under the name *tagoto-no-tsuki*: "(reflected) moon in (the water surface of) every paddy field" (AGENCY FOR CULTURAL AFFAIRS, 2003).

In 1999, the Ministry of Agriculture, Forestry and Fisheries designated the most beautiful rural cultural landscapes of Japan. Obasute *tagoto-no-tsuki* is of course among them. With such a designation, the civic awareness of the unit "scenic beauty place and its surrounding environments" should and could be increased and the disposition for cultural landscape conservation projects enhanced (AGENCY FOR CULTURAL AFFAIRS, 2003). Places of natural beauty, similarly to special species, operate as an incentive for citizens, to get involved in cultural landscape conservation activities. A very vivid example of this is the "*tanada* ownership system", where people from the city rent a piece of rice terrace and cultivate it under the guidance of the landowner and/or local people (KIENINGER, PENKER & YAMAJI, in preparation).

7 The fun-factor and civic involvement in nature conservation activities

The Japanese government explicitly wishes for increased civic engagement in biodiversity conservation activities including nature restoration and conservation of the *satoch*[3]-*satoyama* areas (GLOBAL BIODIVERSITY STRATEGY OFFICE, 2007). The slogan "It's summer vacation, let's catch insects!" in the pamphlet "Living with Nature – The National Biodiversity Strategy of Japan" of the NATURE CON-

SERVATION BUREAU (2002) of the Japanese Ministry of Environment demonstrates how civic involvement is stimulated[4]. In the Japanese way of nature conservation the fun-factor plays a very important role. Humans are regarded as a part of nature and therefore nature conservation is not an abstract concept, but a concrete and participative active process. Nature conservation comes from the heart, and nature is not protected just for its own sake, but for humans too. They protect what they love.

Japanese people, for instance, love dragon-flies. The esteem for these animals has an old tradition in Japan. The Japanese love them as motives in paintings and in poetry, they love to replicate them in children's toys, they love to watch and identify them and they also love to hunt them. Catching dragonflies is an old children game still played today and different catching techniques have been developed. The Japanese public is very interested in protecting dragonflies precisely because they are so well loved. In summary usage and protection belong together.

The "Dragonfly pond concept" is an interesting example: Japan has 180 species of dragonflies; due to their habitat destruction, 41 species are already considered rare or endangered (PRIMACK et al., 2000). Their loss would not only be a biological one, but, as already mentioned, also a cultural one. The dragonfly pond restoration project started 1986 in Yokohama and spread out all around the country. According to PRIMACK et al. (2000), presently more than 500 dragonfly pond projects exist in Japan, with the goal, to restore natural or artificial ponds and to construct new ones. The dragonfly pond projects are organized by local governments and citizens together and enjoy great popularity, especially in urban areas. Dragonfly pond projects are often used in schools as enlarged outdoor classrooms and different subjects such as art, chemistry, botany, plant morphology, ecology, zoology etc. are teached with them: "*Although the popular focus is mainly on dragonflies, the results of grassroots action by an interested public are being felt by entire aquatic ecosystems*" (PRIMACK et al., 2000).

8 Conclusions

The understanding of nature and the appreciation of its objects and phenomena is part of a society's conception of the world. In Europe (and the parts of the world settled by Europeans) nature is viewed as separate from humans, something outside of man. This leads to conservation concepts and programs aiming to separate man and nature; landscapes are segregated into natural and unnatural ones. The human population is segregated into preservers of nature (which are mostly not those who own the land), destroyers of nature (often those who own or utilize the land) and the naïve and incapable majority. Quite contrarily in Japan, although modern conservation politics have been imported together with the corresponding western ideology, the resulting activities are more relaxed. The Japanese people's conception of nature does not exclude what has been shaped by human activities but holds "man-made nature" in particular high esteem. The relations of Japanese to their (!) nature are strongly emotional and rather flexible. In Japan quite contrarily to Europe, integrative and participatory approaches to nature conservation are the rule. On the other hand, conservation stakeholders are in a difficult position in Japan, because human interference with nature, even in areas that have been particularly preserved for nature, is not seen as such a serious offence as it is in Europe.

Notes

[1] "*That was an explicit political event, ...*" JANZEN about the Forum on Biodiversity in Washington DC in 1986, in TAKACS (1996).

[2] "*Although at first blush an apparently more 'scientific' term than wilderness, biological diversity in fact invokes many of the same sacred values ...*" (CRONON, 1995).

[3] The word satochi (sato = village and chi = soil/ground/ living place) is less as popular as satoyama. It stands for rural landscape, including satoyama, farmland, settlement and reservoirs, while satoyama, in the context of satochi-satoyama, just indicates coppice woodlands, pine forests and grasslands (TAKEUCHI, 2001).

[4] Such an appeal by a nature conservation agency to catch insects is unimaginable in Europe.

References

AGENCY FOR CULTURAL AFFAIRS (2003): The Report of the Study on the Protection of Cultural Landscapes Associated with Agriculture, Forestry and Fisheries. (Agency for Cultural Affairs, Japan, Cultural Properties Department, Monuments and Sites Division, Committee on the Preservation, Development and Utilization of Cultural Landscapes Associated with Agriculture, Forestry and Fisheries. Retrieved on July 21, 2008 from ⟨http://www.bunka.go.jp/english/pdf/nourinsuisan.pdf⟩.

COUNCIL OF EUROPE PUBLICATION (1996): Biodiversity: Questions and Answers (Brochure published by the EU Brussels).

CRONON, W.J. (1995): The trouble with wilderness, or getting back to the wrong nature. In: CRONON, W.J. (ed.): Uncommon Ground: Toward reinventing nature. W.W. Norton & Company, New York.

ESER, U. (2001): Die Grenze zwischen Wissenschaft und Gesellschaft neu definieren: boundary work am Beispiel des Biodiversitätsbegriffs. In: Verhandlungen zur Geschichte und Theorie der Biologie, Bd. 7, Berlin VWB 2001, 135–152.

ESER, U. (2003): Der Wert der Vielfalt: „Biodiversität" zwischen Wissenschaft, Politik und Ethik. In: BOBBERT, M., DÜWELL, M., JAX, K. (eds.): Umwelt – Ethik – Recht. Francke Verlag, Tübingen. 160–181.

FUJIOKA, M., H. YOSHIDA (2001): The Potential and Problems of Agricultural Ecosystems for Birds in Japan. Global Environ. Res. 5 (2), 151–161.

GLOBAL BIODIVERSITY STRATEGY OFFICE (2007): The Third National Biodiversity Strategy of Japan. Outline of The Third National Biodiversity Strategy of Japan. (Global Biodiversity Strategy Office, Nature Conservation Bureau, Ministry of the Environment. Retrieved on July 7, 2008 from ⟨http://www.biodic.go.jp/convention/The%20Third%20NBS.pdf⟩.

HARMON, D. (2002): Preserving cultural and linguistic diversity can help protect biological diversity. In: DUDLEY, W. (ed.): Biodiversity. Current Controversies. Greenhaven Press, San Diego, CA, 205–212.

HEYWOOD, V.H. (1994): The measurement of Biodiversity an the Politics of Implementation. In: P.L. FOREY, C.J. HUMPHRIES and R.I. VANE-RIGHT (eds.): Systematics and Conservation Evaluation. Systematics Association, Special Volume No. 50. Clarendon press, Oxford, 15–22.

HOFFMANN, A., S. HOFFMANN, J. WEIMANN (2005): Irrfahrt Biodiversität. Eine kritische Sicht auf europäische Biodiversitätspolitik. Metropolis-Verlag, Marburg.

ICHINOSE, T. (2007): Restoration and conservation of aquatic habitats in agricultural landscapes of Japan. Presentation held on the Workshops and Symposia "Biodiversity and Sustainable Development – Ecological and Socio-Economic Challenges for the Conservation and Restoration of Wetlands in Japan and Europe", at the Bavarian Academy for Nature Conservation and Landscape Management (ANL) Laufen, Germany, on July 2, 2007.

KIENINGER, P., W. HOLZNER (in press): The Impact of ancient "Man/Nature" Concepts on Contemporary Nature Conservation Attitudes in Europe and Japan. ECO-THEE-2008, 2–6 June 2008, Conference, Kolymbari/Crete, Conference Proceedings.

KIENINGER, P., M. PENKER AND E. YAMAJI (in preparation): Townspeople as Paddy Farmers – Cultural Landscape Conservation in Japan.

KOBORI, H., R.B. PRIMACK (2003): Participatory Conservation Approaches for Satoyama, the Traditional Forest and Agricultural Landscape of Japan. Ambio 32 (4), 307–311.

KONISHI, M. (2004): Japan, Land der Libellen. In: ISHIKAWA, J.(ed.): Nipponia. Japan entdecken. Sonderbeitrag* Japan, Land der Langlebigen. Nr. 29, Heibonsha, Bunkyo-ku, Tokyo, 25.

KURECHI, M., S. IWABUCHI (2005): Fuyumizutanbo, environmentally sound rice field. Winter-flooded Rice Fields. Japanese Association for Wild Geese Protection. Tohoku regional nature protection center for the Ministry of the Environment Japan.

KURITA, H., T. MINETA, K. ISHIDA, T. ASHIDA, H. YAGI (2006): Environmental Potentials of Winter-flooded Rice Fields for the Wetlands Restoration. Journal of the Japanese Society of Irrigation, Drainage and Reclamation Engineering 74 (8), 713–717. (in Japanese)

LAWTON, J.H., J.R. PRENDERGAST, B.C. EVERSHAM (1994): The numbers and spatial distributions of species: analyses of British data. In: P.L. FOREY, C.J. HUMPHRIES, R.I. VANE-RIGHT (eds.): Systematics and Conservation Evaluation. Systematics Association, Special Volume No. 50. Clarendon press, Oxford, 177–195.

MINISTRY OF AGRICULTURE, FORESTRY AND FISHERIES, JAPAN (2005): MAFF UPDATE. Implementation of Direct payment to farmers in the hilly and mountainous areas in Japan, 2004. (Retrieved on July 17, 2008 from ⟨http://www.maff.go.jp/mud/595.pdf⟩.

MIYAURA, T. (this volume): Satoyama – a place for preservation of biodiversity and environmental education.

NATURE CONSERVATION BUREAU, MINISTRY OF THE ENVIRONMENT, GOVERNMENT OF JAPAN (2002): Living with Nature. The National Biodiversity Strategy of Japan, Tokyo.

NORTON, B.G. (1994): On what we should save: the role of culture in determining conservation targets. In: P.L. FOREY, C.J. HUMPHRIES, R.I. VANE-RIGHT (eds.): Systematics and Conservation Evaluation. Systematics Association, Special Volume No. 50. Clarendon press, Oxford. 23–39.

OHSAWA, M., T. KITAZAWA (this volume): Biocultural diversity and functional integrity of Japan's rural landscape.

PRIMACK, R.B., H. KOBORI, S. MORI (2000): Dragonfly Pond Restoration Promotes Conservation Awareness in Japan. Conservation Biology 14 (5), 1553–1554.

TAKACS, D. (1996): The idea of biodiversity. Philosophies of paradise. The John Hopkins University Press, Baltimore and London.

TAKEDA, M., T. AMANO, K. KATOH, H. HIGUCHI (2006): The habitat requirement of the Genji–firely Luciola cruciata (Coleoptera : Lampyridae), a representative endemic species of Japanese rural landscapes. Biodiversity and Conservation 15, 191–203.

TAKEUCHI, K. (2001): Nature Conservation Strategies for the 'SATOYAMA' and 'SATOCHI', Habitats for Secondary Nature in Japan. Global Environ. Res. 5 (2), 193–198.

WATANABE, T. (2003): Present Problems of Hilly and Mountainous Areas under Enforcement of the Direct Payment System. In: PRIMAFF Annual Report 2003. (Retrieved on July 17, 2008 from ‹http://www.maff.go.jp/primaff/koho/seika/annual/pdf/annual2003/an2003-6-7.pdf›).

WEIMANN, J., A. HOFFMANN, S. HOFFMANN (2003, eds.): Messung und ökonomische Bewertung von Biodiversität: Mission impossible? Metropolis-Verlag, Marburg.

Adress of authors

Dipl.-Ing. Pia Kieninger, Univ.-Prof. Dr. Wolfgang Holzner, Dr. Monika Kriechbaum, Center for Environmental Studies and Nature Conservation, Department of Integrative Biology and Biodiversity Research, University of Natural Resources and Applied Life Sciences, Vienna, Gregor Mendel Straße 33, 1180 Vienna, Austria.

Eingelangt am 10. September 2008
Angenommen am 7. November 2008

Artikel 2: Kieninger, P.R., Yamaji, E., Penker, M. 2011. Urban people as paddy farmers: The Japanese Tanada Ownership System discussed from a European perspective. Renewable Agriculture and Food Systems 26 (4): 328–341.

Urban people as paddy farmers: The Japanese Tanada Ownership System discussed from a European perspective

Pia R. Kieninger[1]*, Eiji Yamaji[2], and Marianne Penker[3]

[1]Institute of Integrative Nature Conservation Research, Department of Integrative Biology and Biodiversity Research, University of Natural Resources and Life Sciences, Vienna, Gregor Mendel Strasse 33, 1180 Vienna, Austria.
[2]Department of International Studies, Division of Environmental Studies, Graduate School of Frontier Sciences, The University of Tokyo, 5-1-5 Kashiwanoha, Kashiwa City 277-8653, Japan.
[3]Institute of Sustainable Economic Development, Department of Economics and Social Sciences, University of Natural Resources and Life Sciences, Vienna, Feistmantelstrasse 4, 1180 Vienna, Austria.
*Corresponding author: pia.kieninger@boku.ac.at

Accepted 28 February 2011; First published online 15 April 2011   Research Paper

Abstract
The degradation of the traditional cultural landscape due to abandonment of agricultural management is perceived as a serious problem in different parts of the world. Rising consciousness concerning this issue in Japan led to the formation of numerous voluntary civil farming programs. This paper presents a multi-method case study conducted in Japan (Ōyamasenmaida, Chiba prefecture) about a highly relevant rural–urban cooperation, where landholders lease out their rice terraces to city dwellers to grow their own rice under the intensive instruction and well-organized support by local farmers and other local experts. The activity is known as 'Tanada Ownership' (tanada means rice terrace). It is spread over the country and promises, in contrast to the short-term individualistic European models, long-term rural–urban relations, the valorization of local knowledge and natural resources as well as the maintenance of the rice-terrace landscapes in several regions of Japan. The particular goal of this research is the investigation of the 'how' and 'why' of the system and the motivation of its participants. The Japanese approach is compared with similar European initiatives and conclusions focus on the particularities of the Japanese Ownership System and its transferability to the European context. Despite the innovativeness and popularity of the Ownership System, scientific studies are relatively scarce and none of them published are in English. This article therefore presents an original and important contribution to the scientific community, as it provides insights into the Tanada Ownership System and puts it into an international context by comparing it with European initiatives of voluntary farm work.

Key words: Japan, cultural landscape conservation, Ōyamasenmaida (Chiba Prefecture, Kamogawa City, Japan), Tanada Ownership System, voluntary civic engagement, voluntary farm work

Introduction

Voluntary farm work of the urban population had some tradition in Europe during and after the Second World War. Urban volunteers attended (school) harvest camps, week-end farm clubs and other land-based activities to mitigate food shortages[1,2]. In the following period of surplus food production, such engagements became irrelevant. Recently, not only in Europe but also in other industrialized countries, volunteers and civil society organizations started to support direct involvement with land and farm work again[3,4]. In the

USA, the term 'civic agriculture' was created in 1999 by T.A. Lyson to highlight small(er)-scale (family) farms and local-community-based food production activities and associations as counter movement to the US–American industrialized, globalized, large-scale agriculture[5]. Civic agriculture is a very broad 'concept' characterized by local rural–urban networks between consumers and producers striving for an economically, environmentally and socially sustainable agricultural system, including, e.g., organic farming, direct marketing, alternative (agri)-food stores, and/or movements[5]. Also 'urban agriculture', 'community gardens' (in Europe 'allotment gardens') or the 'farm to school (FTS) program' can be defined as civic agriculture[6–8]. In contrast to the Japanese Tanada Ownership System presented in this article, urban agriculture and community gardens take place in an urban context and mostly on public land. The FTS program aims at prevention of child obesity by including nutrition in the curriculum and by changes in the school lunch menus toward more fresh food (fruits, vegetables, eggs, honey, meat, etc.) from local farmers[7]. Supporting local farmers helps to revitalize the rural community[7]. In contrast to the systems analyzed in this paper, a direct involvement of teachers and pupils in agricultural activities, beyond farm visits, however, is not common[7,9].

This paper presents a popular, well-established and organized form of direct civic voluntary farm work in the Japanese rice terrace landscapes that are endangered by abandonment of agricultural land use. The results are discussed against literature on voluntary farm work, landscape stewardship and nature conservation in Europe, Australia and the USA. Conclusions regarding the particularities of the Japanese Ownership System and its transferability to the Western context are drawn.

Rice is the traditional main staple food of Japan. Its cultivation has a tradition of more than 2000 years. It can therefore be considered as a fundamental feature of the Japanese cultural landscape, which is a place of high biocultural diversity[10]. Following the Western lifestyle the eating habits of the Japanese people have changed and rice consumption decreased by half from 118.3 kg to an annual per capita consumption of 59 kg between 1962 and 2008[11]. Furthermore, because of a very small average farm size (1.38 ha[12]) and high costs of labor[13], Japanese rice farmers are confronted with up to 10 times higher production prices, compared to foreign producers[14]. From an economical point of view, the cultivation of rice therefore becomes more and more unalluring. Thus, it is not surprising that the number of farm households has decreased by 59% between 1950 and 2008[11,12]. The farmland area diminished by 24% (from 6.09 million hectares in 1961 to 4.63 million hectares in 2008) to 12% of the country's territory[12]. A total of 59% of the agricultural key workers were 65 years or older in 2008[12]. These structural and demographic developments are reflected not only in a continuously dwindling food self-sufficiency rate to the lowest among the major developed countries[15] but also in the loss of biocultural diversity associated with traditional agricultural landscapes.

The traditional cultural landscape of Japan has been shaped by thousand of years of agricultural land use[10]. Because of its long history and the co-evolution of natural and social systems, the landscape is very rich in different habitat types, plant and animal species, customs and culture. The Japanese people perceive it as their 'mother landscape'[16], as some kind of arcadia, they are longing for. Its degradation due to an insufficient level of management is listed in the Fourth National Biodiversity Strategy as one of the main reasons for the loss of biodiversity[17].

In the late 1980s, the first cultural landscape conservation movements were emerging at the local level[18]. Whereas engagement in woodland management is restricted only by few legal rules, there were many legal limitations for civil society participation in agricultural cultivation[18]. With very few exceptions, exclusively farmers were allowed to own and cultivate agricultural land. Non-farmers could just engage in public allotment gardens or to support existing farmers as volunteers[18]. However, some years later a special agricultural leasing partnership program, called 'ōnā seido', was initiated, where non-farmers have the opportunity to rent agricultural land. Ōnā is the 'Japanized' English word for owner, whereas seido means system. Ōnā seido translated into English makes 'Ownership System'. A more correct term, however, would be 'Tenure System', as the participants in the Ownership System, the owners (ōnā), are not owners, but tenants.

The Ownership System can be regarded as an urban–rural coalition, where non-farmers (predominantly city dwellers) engage in farming activities. Against the payment of a participation fee, they rent their own agricultural land, cultivate it and thus contribute to the revitalization of the rural society and preservation of the traditional cultural landscape.

As it is generally prohibited in Japan by agricultural legislation to lease farm land out to non-farmers, until 2003 the leasing out in the Ownership System could only be done indirectly via a city or an agricultural cooperative association[19]. Later, however, a special law came into action, which allows every prefecture and/or city to apply for the designation as a 'special district for policy renovation' ('tokku'). In these tokku districts, the leasing restrictions do not apply and farmers are allowed to lease out their agricultural land directly to non-farmers[19].

Apart from urban participants (tenants), landowners and local supporter groups are further stakeholders in the Ownership Programs. The local supporter groups play a fundamental role because the landowners are often too old for field work, while the mostly inexperienced tenants need guidance. Moreover, tenants participate only in some special occasions a year, as for example, sowing, transplanting rice, weeding or harvesting. All the work in between is carried out by either the landowners or the local supporter groups.

The Ownership System became very popular in Japan and spread throughout the country, in particular the Tanada Ownership System. Tanada is the Japanese word for rice terrace/terraced paddy fields. In most cases, rice terraces are situated in the mountainous regions of Japan, which are most threatened by land abandonment[20]. But as also shown in the study of Swanwick[4], people prefer landscape with diverse landforms to lowlands. So the popularity of Ownership Programs in rice-terrace landscapes may also be explained with their aesthetic beauty[21,22]. Apart from that, they evoke strong feelings of timelessness, identity and links with the past[4].

The first Tanada Ownership System started 1992 in the village Yusuhara (Kochi Prefecture, Japan) in Shikoku

Island[23]. All around Japan 187 Ownership programs were counted in 2008[24].

Compared to the popularity and innovativeness of the Ownership System, scientific studies are relatively scarce[22,25–27]. Besides that, all of them have been published exclusively in Japanese. Thus, there is little knowledge of this movement among non-Japanese scientists. The particular goal and groundbreaking aspect of our research is to investigate the system from a European viewpoint and to provide a new and deep understanding in the Ownership System by the holistic research approach. The Japanese Ownership System is compared with similar European initiatives. Conclusions focus on the particularities of the Japanese Ownership System and its transferability to the European context.

The factors determining the choice of the study site were the acknowledged scenic and ecological value of its cultural landscape and personal contacts with the local population.

Prior to this study, some research was done on vegetation[28], visitor's perception of the scenery of tanada landscapes[29], cultural landscape conservation methods[30] and on the Ownership System itself[26,27].

The questions guiding the present research are of clear exploratory nature:

- Why did landowners and local people found the Ownership System in Ōyamasenmaida?
- How is the Ownership System organized?
- What are the provenance, professional background, age and motivations of the tenants of the Ownership System?
- What are the tenants' attitudes regarding cultural landscapes and different ways of protecting it?

The study area

Ōyamasenmaida is a rice-terrace landscape located in the mountains (Fig. 1), in the south-eastern end of the Boso Peninsula (Chiba Prefecture, Kamogawa City, Japan), around 100 km to the south-east of Tokyo. The name Ōyama (big [ō] mountain [yama]) is derived from Mt. Ōyama (with a shrine on its peak) and Senmaida stands for thousand [sen] pieces [mai] of paddy fields [da]: 'Thousand pieces of paddy fields located in the mountainous area'.

Next hamlet to Ōyamasenmaida is Kogane (Fig. 1, left upper corner) with fewer than 20 households. Kogane is part of Hiratsuka (165 households, 532 inhabitants in 2002), which is one of the six villages situated around Mt. Ōyama[31]. The terraces extend on a south-east slope over an elevation difference from 80 to 150 m above sea level and vary in their size between 20 and 900 m^2.[30] The rice-terrace complex of Ōyamasenmaida counts 375 rice terraces on an area of 3.5 ha; the Ōyamasenmaida Preservation Association (PA) covers adjacent fields and operates 415 plots (4.5 ha)[30].

Around two-thirds of the terraces are cultivated with rice (one-half each landowner's and Ownership System) (unpublished data). The remaining third is abandoned or rather a small part is used as orchard.

In 1999, the tanada landscape of Ōyamasenmaida became designated by the Japanese Ministry of Agriculture, Forestry and Fisheries as one of the 'Top 100 Terraced Paddy Fields of Japan'[23]. The policy behind this nomination was to direct the population's attention to the rice-terrace landscapes of Japan[32].

Methods

The field research for this paper was conducted over 50 days from December 2004 to May 2006.

As the research topic has exploratory character and concerns a contemporary phenomenon within its real-life context without distinct boundaries between context and phenomenon, a case study approach was chosen[33]. The methods adopted for data collection were participatory observation (PO)[34], personal communication (PC), semi-structured expert interview (EI) and questionnaire-based survey (QS).

PO and PC

Between December 2004 and May 2006, the first author participated in different activities in Ōyamasenmaida: e.g., assembly of the PA, several tenants, meetings, rice planting, seasonal traditional cultural events or ordinary workday life of the PA staff and local people. In the PO process (which includes all sense)[34], daily conversation with tenants (tenants A, B, D, F), supporters, staff members and land owners was a very important way for information gathering. Major support came from the PA staff members and main supporters. Complex topics of conversations were accompanied by interpreters, such as English-speaking research colleagues, tenants, supporters or members of the PA.

Semi-structured EI

About 6 months after the start of the field work in June 2005, the director of the PA, who is also the head of the Ownership System, was interviewed. The face-to-face EI was based on semi-structured interview guidelines. A Japanese PhD student who studied the vegetation of Ōyamasenmaida helped with the interview as interpreter. In November 2005 and January 2006, a long-lasting tanada ownership tenant (tenant A) was interviewed twice in a semi-structured EI.

Questionnaire-based survey

Based on the experiences of interviews and PO, structured questionnaires were designed in order to gather more detailed information regarding the PA, the Ownership System and its members. Target groups of the questionnaires were

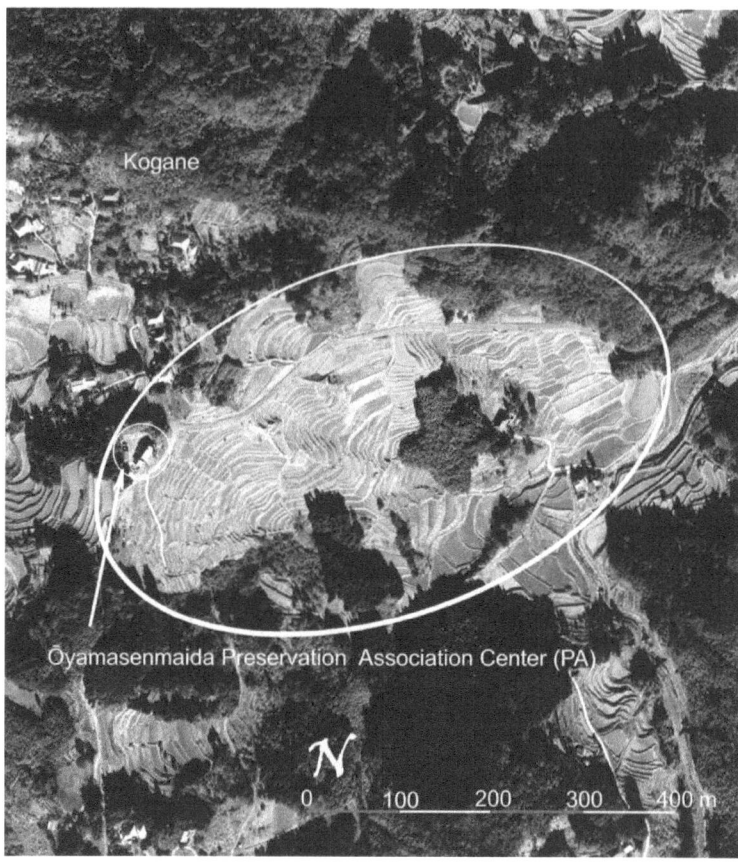

Figure 1. Aerial photo of the area of Ōyamasenmaida, taken on January 15, 2004; source: Keiyō Survey Co.

the eight landowners of the rice terraces, the 30 main supporters of the Ownership System and the 453 tenants.

For the questionnaire, the paper form was chosen, since e-mail addresses were not known in most of the cases. The structured questionnaire included behavioral, attitudinal and classification questions, which were asked in open, semi-closed and closed style.

The questionnaires were adapted for each of the three groups and distributed in spring 2006. They had been written in English and translated afterwards into Japanese. The open answers in the questionnaires had to be translated back from Japanese to English.

The cover letter gave a short introduction and explained the reason for the survey. With some additional sentences, the head of the Ownership System motivated the tenants to participate.

Tenants. Participation lists and addresses were provided by the PA office. The questionnaires (complete inventory count) were distributed out personally in tenant meetings or sent out (together with a stamped reply envelope) to those not attending. The response rate was 55% (see Fig. 2).

Landowners and main supporters. The questionnaires were personally handed out at the general meeting of the PA. To those not attending, the questionnaire was delivered personally by the PA staff or sent by post (together with a stamped reply envelope). Response rate was 38% and for the main supporters 37% (see Fig. 2).

The results from the questionnaire surveys, the EIs, observation notes and PC in field, as well as the information gathered from up-to-date documents, web sites, official statistics and secondary data, were triangulated for

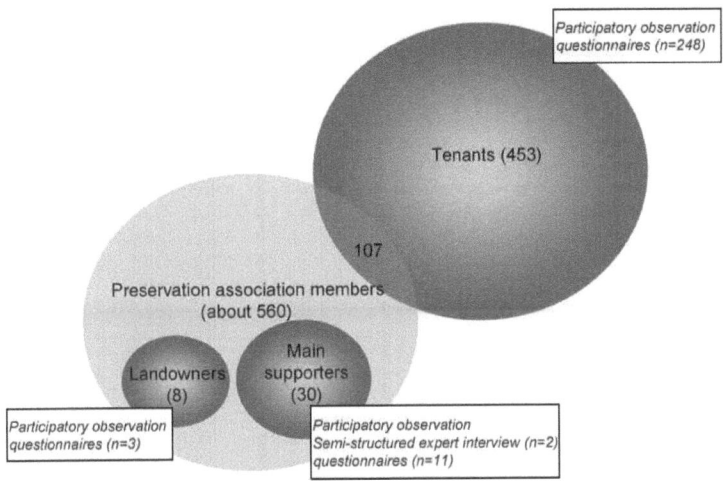

Figure 2. Members of the PA and the Ownership System, methods of data collection and return rate of the questionnaires; source: EI: June 18, 2005; PC: August 2, 2009.

validation and if necessary counterchecked by additional e-mail or phone conversation after the field-research period.

Results

Reasons for the foundation of the PA

In 2006, the rice terraces of Ōyamasenmaida belonged to 11 proprietors, all of them part-time farmers (PC: director, June 18, 2005). While three of them were still of working age, the remaining eight had already retired (ibid.). In 1997, these eight senior landowners founded, together with other farmers and local people, the NPO 'Ōyamasenmaida Preservation Association', as for all of them, these rice terraces were something special which they wanted to preserve from abandonment and decay: '*We hope, that the landowner's children will continue the husbandry of their parents in their retirement. But even if they won't do so, it is not a problem. The system is also survivable without farmers, because the management of the rice terraces is done by the Ōyamasenmaida Preservation Association group*' (PC: director, June 18, 2005).

After having established the PA, their members decided that an Ownership System, which they had heard about being applied in other regions in Japan, would be also an ideal instrument for Ōyamasenmaida (PC: director, June 18, 2005). Therefore, they initiated in 1997/1998 the 'Ōyamasenmaida Tanada Ownership System': '*The area of Ōyamasenmaida has many disadvantages, as a strong depopulation, very small scale uneconomical paddy fields on a steep inclination and a highly aged population. These factors can't give much power to the region. So we thought about, how we can change these disadvantages into advantages. We think that a Tanada Ownership System is the right way for the paddy, because the small size of the paddy fields is still big enough for city dwellers to cultivate it. The old farmers, even if they may not have so much power any more, they however posses a lot of knowledge to offer to the city people. In contrast, city people have a big lack of nature, pure and clean air. If we add up both disadvantages – that one of the city dwellers and that one of Ōyamasenmaida – we gain an advantage*' (ibid.).

As the tokku districts were not already established at this time, the lease-out process had to be carried out via a city (EI: director, June 18, 2005; PC: director, September 26, 2009). Ten percent of the participation fee was therefore retained by Kamogawa City for administrative tasks, 10% went to the landowners and 80% to the PA (ibid.). The contract was set until 2005 (PC: director, September 26, 2009). In 2003, Kamogawa City applied to the Japanese Ministry of Agriculture, Forestry and Fisheries to become a 'Rice Terrace Agriculture Tokku' (EI: director, June 18, 2005; PC: director, January 28, 2010). The petition was accepted in 2004 and Ōyamasenmaida started thereupon with a second Ownership System, the 'Kamogawa City Ōyamasenmaida Tanada Ownership System' (PC: director, January 28, 2010). In this new Ownership System, the PA is allowed to directly lease out their land without any loop via Kamogawa City (EI: director, June 18, 2005). About 10% of the rent goes to the landowners, the rest (90%) to the PA (ibid.). When the contract of the 'Ōyamasenmaida Tanada Ownership System' expired in 2005, it was not renewed (PC: director, September 26, 2009). The old Ownership

System passed into the new one; the two different names, however, are still in use (ibid.).

Being a member of the PA is independent from the Ownership System (PC: director, August 1, 2009). But in more than 100 cases, tenants are also PA members (Fig. 2).

With 'Ōyamasenmaida' the PA wants to show the 'ideal model' of a cultural landscape and demonstrate ways how to preserve mountainous areas (EI: director, June 18, 2005). For the PA, at least as important as planting, harvesting, mowing, etc., is the exchange and the communication among the participants themselves and between the tenants and the local people (ibid.). Therefore, social gatherings are very important (PO: e.g., karaoke evenings after farming activities 2005/2006, harvest festival October 2005). Townspeople should be aware of the problems the mountainous areas are confronted with, and they should take it seriously, because: '*Most of the Japanese population lives in cities. The government is in the city. Big and important decisions are made by city people in the cities. City people can change the government and they make money; money, which is necessary to protect the mountainous areas*' (EI: director, June 18, 2005). In order to facilitate and encourage the rural–urban communication, the 'Tanada Club house' was erected in 2000, financed by the Kamogawa City (ibid.).

In the area of Kamogawa City, besides the Ownership System in Ōyamasenmaida, six other Ownership Systems exist[35].

Budget of the PA

The budget of the PA results from the yearly membership fees, the Ownership System fees, (see below) and the attendance fee of other seasonal activities/participatory programs. Besides that, Ōyamasenmaida-related products, such as rice, soybean paste, calendars, postcards, charcoal, etc., are sold (PO: December 2005–June 2006, November 6–8, 2010). During voluntary farming days or festivals, local dishes and drinks (e.g., dandelion coffee, curry rice, o-bento [Japanese lunch] with rice balls, miso soup and pickles) are prepared by female members of the PA and sold at moderate prices (PO: working events 2005/2006, torch festival November 6–8, 2010). In addition, the seminar room of the 'Tanada Club' is rented out for meetings, workshops, etc., at the price of 400¥ (c. 3.5€) h^{-1}.[36] (The conversion from Yen to Euro of all prices in this paper is based on the foreign exchange rate of the Bank Austria, June 22, 2010.) Finally, the PA receives some donation from private companies, for example, from the CSR (Corporate Social Responsibility).

From these incomes office expenses, ongoing maintenance measures, permanent staff (three people), landowners and instructors are paid. All other people (around 70 people, 30 of them can be regarded as main supporters) are working as volunteers or are paid 800¥ (c. 7€) h^{-1} (EI: tenant A, January 20, 2006).

Structure and organization of the Ownership System

In 2006, the tenants of the Ownership System cultivated most of the land of the eight retired farmers (EI: director, June 18, 2005). The above-mentioned three younger landowners, however, did not allocate their land, as they still cultivated it themselves (ibid). After the field research period, three of the eight senior farmers passed away (PC: director, January 28, 2010). Their land was inherited by their children and is still part of the Ownership System. Out of the three non participating landowners, one entrusted his land to the PA, while the second sold it to the third who is still an active farmer (PC: director, January 28, 2010).

The PA offers five different ownership programs[37]:

(1) The oldest program is the Tanada Ownership System[37]. It started in 2000 as 'Ōyamasenmaida Tanada Ownership System'[37]. From initially 39 rented rice terraces, the number increased to 136 in 2002 (PC: director, December 8, 2009). The tenants of the Tanada Ownership System have not changed since that time (EI: tenant A, November 5, 2005). Every rice terrace has one registered tenant—even if he/she is participating with family, friends, colleagues, students, etc. (PO: rice planting events April/May 2005/2006). The participation fee is 30,000–40,000¥ (262–340 €) yr^{-1} per 100 m^2,[37] adjusted in proportion to the area (PC: tenant A, June 29, 2010 and tenant B, March 14, 2010). The harvested rice of all tenants is put together and then divided pro-rata (PC: tenant B, March 14, 2010).

(2) In the Soybean Trust System, a special program of the PA, which started in 2001, soybeans were cultivated on abandoned rice terraces in order to revitalize them (EI: director, June 18, 2005). In contrast to the Tanada Ownership System, tenants do not rent their own 'soy terraces', but shared them, as the focus of this program is the working event *per se*[37]. The participation fee is 4000¥ (35 €) yr^{-1} participant^{-1}.[37]

(3) The Tanada Trust System is a structure equivalent to the Soybean Trust (collective group rent of a rice terrace) and has existed since 2002[37]. The participation fee amounts to 30,000¥ (262 €) yr^{-1} per 100 m^2.[37] The harvested rice is divided equally; around 30 kg to each registered participant[37]. In comparison with the tenants of the Tanada Ownership System, the Tanada Trust members fluctuate more and their number changes every year (PC: director, January 12, 2007).

(4) The Rice Wine Ownership System was launched in 2004[37]. Despite its name 'Ownership System', it is rather a Rice Wine Trust System, as the tenants collectively grow rice on common paddy fields (PC: tenant B, March 9, 2010). After the harvest the rice of all Rice Wine tenants is delivered to a traditional rice wine brewery in the vicinity (PC: tenant B, December 22, 2009). Every tenant gains three 1.8-liter bottles of wine[37]. Included in the Rice Wine Ownership System, traditional rice wine drinking cups and rice wine labels

(determined by a competition) are produced[37]. The participation fee is 15,000¥ (131 €)[37].

(5) The most recent program (since 2005) is the Indigo–Cotton Trust System[37]. On an area about 2 km from Ōyamasenmaida cotton and indigo are collectively cultivated (PO: October 15, 2005). After the harvest, the cotton is yarned and clothes are woven and later dyed with indigo (PC: tenant F, spring 2006).

Apart from these five programs, other activities are offered as well, e.g., lectures on fireflies and firefly watching (PO: June 2005), bamboo torch events around the rice terraces (PC: director, January 10, 2009; tenant B, January 24, 2009; tenant D, January 18, 2008; PO: November 6–8, 2010), concerts with traditional dances performed on the rice terraces[37] [PO: Kagura performance (theatrical dance), November 7, 2010], photo award of Ōyamasenmaida for the yearly calendar (PO: contest participation, July 7, 2006), production of traditional Japanese handcrafts and dishes (PC: tenant B, December 22, 2009; PO: rice planting events April/May 2006; hand craft events October 15, 2005, December 18, 2005), volleyball tournaments in the muddy paddy fields before rice planting (PO: April/May 2006), etc.

From the number of participants and size of area, the strongest programs are the Rice Wine and the Tanada Ownership System. Indigo–Cotton and Soybean are less important.

Many participants attend more programs contemporaneously. That is why the questionnaire was sent to all participants. In the empirical inquiry, however, the focus of PO and PC was concentrated on the rice-growing tenants (i.e., Tanada Ownership System, Tanada Trust System and Rice Wine Ownership System).

Yearly schedule of a Tanada Ownership and Tanada Trust System tenant

For the 'tanada tenants', farming activities are scheduled about seven times a year, mostly with two time options: rice planting in April/May, weeding in June, July and August, harvesting and threshing in September and harvest festival in October (EI: tenant A, November 2, 2005). All other work in between (mowing or burning of the slopes, manuring, sputtering, etc.) are done by the landowners or the PA (EI: director, June 18, 2005 and tenant A, November 2, 2005).

The seven collective working days are big events and are very accurately organized. Each day starts with an attendance check at the registration desk and a welcome speech to explain the respective procedure (PO: April/May 2005/2006). Most of the tenants come by car in the morning and return the same day (PO: working events 2005/2006). The tenants who stay overnight mostly spend the night in farmers houses for a small payment or for free, depending on how well they know each other: '*I always stay at [...]-san, but he is not the landowner of our [...] rice field. Now he does not work for tanada club. He invited me for omatsuri [folk festival] in private. This time, I did not go to the omatsuri but I stayed at his house and had a party. I introduced my friend, who works for a fertilizer company. [...]-san likes to meet people, so not everybody, but who are friends of friends, are welcome to visit his house for a stay. About payment, we always have party, eating and drinking, so I usually pay some money for dinner, like 3,000–4,000¥. He is an old man, living on pension, so that is reasonable. Anyway, this has nothing to do with tanada club. But many people meet these farmers at the tanada club and become personal friends. I guess this is also one aim of tanada club*' (PC: tenant B, August 18, 2009).

The landowners and supporters of the PA instruct the less experienced tenants (PO: working events 2005/2006). Furthermore learning from other tenants plays a role: '*For example, the landowner of my paddy is Mr. [...]. He taught me in the first year how to work. Now I can teach my neighbors*' (EI: tenant A, November 2, 2005). If someone cannot come to do the work on the rice field, the supporters or neighbors will do it, '*but it gives a very bad impression!*' (ibid). For a better identification in the field, the instructors wear red hats with name targets (PO: working events 2005/2006). As the tenants do all labor manually, there is no necessity for special costly equipment or machinery: '*For the work on my paddy field I only bought jikatabi [working shoes with a thick rubber sole and a separate section for the big toe] and a scythe*' (EI: tenant A, November 2, 2005). Cultivated rice in Ōyamasenmaida is mainly koshihikari (EI: tenant A, January 20, 2006), a very popular rice variety in Japan[38,39]. Despite the dominance of manual labor, the farming management in and around Ōyamasenmaida is not organic farming: '*Some of them [farmers] use machines and insert pesticides, chemical fertilizer, herbicides and insecticides, others not*' (EI: tenant A, November 2, 2005).

Provenance and age structure of the tenants, supporters and land owners

The findings presented in the following sections are based on the questionnaire surveys, mostly of the tenants (March–May 2006), but also of the supporters (May–June 2006) and landowners (May–June 2006).

From 248 responding tenants 246 are city dwellers. They live in 55 different cities of five prefectures (two answers could not be attributed) from about 50–150 km distance from Ōyamasenmaida (Fig. 3). The prefectures Tokyo, Kanagawa, Saitama and Chiba, from which most of the tenants come, are densely inhabited (with 1175–5751 persons km^{-2}, among the highest population densities of Japan; PC: Uematsu, M., Population Census Division, Statistic Bureau, MIC, May 11, 2009).

The population size of the 55 cities ranges from over 11,000 up to 8.5 million inhabitants (ibid). The biggest share of the responding tenants (58%) lives in cities with 100,000–1,000,000 inhabitants.

Urban people as paddy farmers

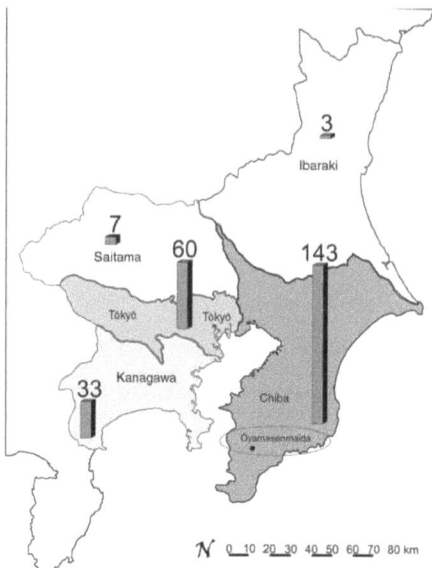

Figure 3. Provenance of the tenants (number of tenants from the relative five prefectures); map based on Bartholomew[40].

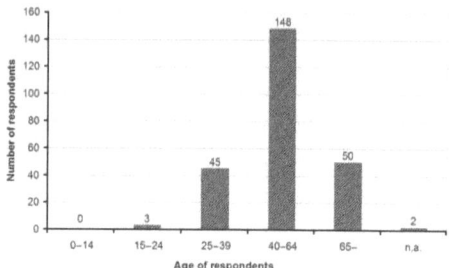

Figure 4. Age structure of the responding tenants (248).

The tenants' questionnaires were answered by 75% male and 25% female tenants. This gender imbalance, however, does not imply that women are underrepresented in the Ownership System but rather that more men than women answered the questionnaire (PO: March 26, 2006).

The average age of the 248 responding tenants was 53 years (two invalid answers). Twenty percent were at least 65 years old or older. Compared with the Japanese population, this share represents the Japanese average age structure (20.1% aged 65 and older)[41]. The class '15–64 years' of the sample comprised 79% (Fig. 4), both over the Japanese average (65.8%) and that of the five prefectures[41]. The age group of 0–14 years accounts for 13.7% for the whole of Japan. In the sample of tenants this group does not appear, as children, even when participating with their parents and/or grandparents, did not answer the questionnaire (PO: working events 2005/2006).

From the 30 main supporters, 11 answered the questionnaire (three women and eight men). The average age was 64 years. The average age of the three landowners (all male), who answered the questionnaire was 76 years.

A big share of the tenants (82%) participated in the company (about three-quarters with family members, one-quarter with friends). Only 17% (43 respondents) stated that they are not accompanied (among them 11 singles and 28 married). In 50% of the accompanied respondents, the participating 'family member' was wife or husband only. The other 50% referred to different combinations of spouse, children, parents, parents-in-law, grandchildren, son, daughter, son- or daughter-in-law.

Tenants' professional background and relation to agriculture

Out of 248 responding tenants, 63% were in employment and 32% were not in employment (49 retirees, 21 housewives, five students and four unemployed persons); 5% of the answers could not be clearly assigned. Among the persons in employment, 150 people (96%) worked in the tertiary sector, 2.6% in the secondary and 1.3% in the primary sector. The proportion of 63% employed persons among the tenants' sample is slightly higher compared to the Japanese average in 2006 (58%)[42]. The share of tenants working in the tertiary sector is significantly higher (96%) in contrast to the Japanese average (67.2%) in 2005[43] and, similarly, the share of tenants of 60 years or older who are still in employment (39 versus 15% Japanese average[42]).

One question addressed in the tenants' questionnaire was: '*How was your relation to agriculture, before attending the activities in Ōyamasenmaida?*' From 251 valid answers (multiple answers were possible; three tenants abstained from answering), 65% stated that they had no connection before, whereas 16% were directly descended from agricultural families or had farmers as relatives (such as parents-in-law, grandparents). In each case, 4% had a farmer as neighbor, visited the countryside at the weekends or grew their own vegetables in kitchen gardens. The remaining 20 answers were very diverse, for example: '*I collect materials about agriculture for my picture books*', '*Produce a musical with the theme on a farming village*', '*When we become 40 years, we would like to become farmers*', '*I worked as a university professor in the field of agriculture for many years*', '*I am alumnus of a Japanese University, Faculty of Agricultural Sciences*'.

Tenants' motivation

When asked for the main reason for their participation in the Ownership System, the most common answer was '*Because I like tanada*' (Fig. 5). This question was asked in

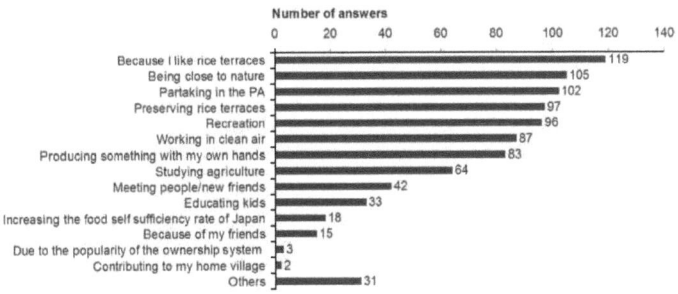

Figure 5. Tenants' motivation for participation (897 valid answers from 237 persons).

a semi-closed style: 14 response options were given and there was also the possibility to add further motivations. The tenants could indicate a maximum of five motivations. From 248 respondents, three abstained from voting and eight selected more than five answers and were therefore rejected.

Among the 31 additional motivations (Fig. 5), nine concerned the interest in rice wine, soybean and/or indigo. The remaining answers were unique, e.g., '*In order to compose haiku*' [Japanese poetry], '*Convalescence from a mental illness*', '*In order to understand Japanese culture through rice cultivation*'.

Relevance of cultural landscapes and ways to maintain it

For 58% of the 248 tenants surveyed, cultural landscape was very important, for 28% important, for 9% less and for 1% not important at all. Nine persons (4%) did not answer this question.

Regarding the question, how to conserve the traditional cultural landscapes, 8% would be willing to pay more taxes, 9% to donate money for farmers once a year and 15% to pay higher prices for farm products. Working as a volunteer once a year (47%) was the most accepted measure to support traditional cultural landscapes. Only 1% of the respondents saw the current landscape development as satisfying.

Discussion

If we try to compare the Japanese Ownership System with similar activities in Europe, North America and Australia (Table 1), we can make out two rough categories of hobbyist, voluntary work on farm land:
(1) 'Self-harvest' projects in several European cities, allowing the urban population to grow their own vegetables on leased farm land.
(2) Volunteers collectively participating in landscape stewardship and nature conservation on farm land.
In the self-harvesting project, well analyzed by Vogl et al.[44], 861 self-harvesters cultivate plots of 20, 40 or 80 m^2 of arable land, prepared and sown by 12 organic farmers in Vienna.

Conservation volunteers are organized on local, national or even international level as civil society movements and work on farm land[45–47]. There are also corporate volunteering and (semi-) commercial organizations placing individual volunteer tourists on organic farms or in nature conservation projects worldwide[48,49].

Common ground between all three forms of voluntary work on farm land

Despite the different geographic contexts and diverging forms of organization, there are astonishing similarities in the motivation. Common to all three forms of voluntary work is that they clearly reflect the shift from food production to consumption of recreation, landscape stewardship, meditation and education (Table 1). In rural geography and sociology, this fundamental pattern of rural change has been discussed under the notion of consumptive or post-productivist countryside[56,57].

Although detailed socio-demographic data are missing, the comparison of the case study results with the literature on self-harvesting and conservation volunteers clearly indicates that the volunteers are not a particular group in terms of socio-demographic characteristics (apart from maybe higher-education level and urban background), but they share similar values and norms, such as the pleasure of socializing, readiness to work manually, and the will to contribute to common goals such as nature or landscape conservation[47,58].

Ownership System versus self-harvesting

Self-harvesting clearly differs from the Ownership System by its individualistic approach, focusing on organic food production, on individual learning and innovations.

The self-harvesters live much closer to the plots (on average 1.8 km[44]) and work on them much more often. They come and go according to their individual schedule, make their own choice of plants and there is even a kind of

Table 1. Comparison of the differences and commonalities among the Ownership System, self-harvesting and conservation volunteering.

Features	Ownership System	Self-harvest initiatives	Volunteers in landscape stewardship and nature conservation
Context	Japan	Austria and Germany[44,51]	Worldwide[45-47,49,50]
Land use	Mainly rice	Vegetables and herbs[44,51]	Grassland, wet lands, orchards, hedges and 'natural' landscapes as, e.g., the rain forest[46,47,50]
Nature of activity	Collective	Individualistic[44,51]	Collective[46,47,49,50]
Main activity	Conventional/traditional food production, preservation of cultural landscape	Organic food production[44,51]	Landscape stewardship, nature restoration, nature monitoring[45-47,49,50]
Motivations of the volunteers	Rice-terrace landscape, being close to nature, recreation, working outdoors, manual labor, studying agriculture, meeting people, educating children (Fig. 5)	Own organic vegetables, resting and meditating, talking with other self-harvesters, picnicking, playing with children, walking around, nature watching, reading a book, sunbathing or taking photographs[44]	Fascination with nature, accomplishing something in a group without hierarchy, physical fitness and health, socializing with like-minded people, slipping into different world, personal growth, living authentically, doing something useful, recreation, learning[45,46,50,52-54]
Fees	262–340 €/100 m² yr⁻¹ for a Tanada Ownership System tenant in Oyamasenmaida[37]	313–375 €/100 m² yr⁻¹ (Austria)[55] 176 €/100 m² yr⁻¹ (Germany)[51]	Voluntary tourists pay agency fees; volunteers in other voluntary organizations do not pay[45,47,49]
Origin of initiative	First (Tanada) Ownership System 1992 in Shikoku; since 2000 in Oyamasenmaida, initiated by the local PA (founded in 1997 by local landowners and citizens)[30]	1987 by an Viennese organic farmer and Regine Bruno from the Environmental Advisory Service, in reaction to the Chernobyl disaster[44,51]	Volunteer Stewardship Network created in 1983; many amateur associations for the pursuit of natural history, founded during the 19th century[45,46,49]
Origin of volunteers	City dwellers from about 50 to 150 km distance	Urban and rural people from a distance of 1.8 km in average[44]	International and local volunteers[45-47,49,50]
Age of volunteers	All age groups	Most between 30 and 50[44]	
Professional background of volunteers	Rather well-educated people (96% of the employed tenants work in the tertiary sector jobs, generally requires higher qualifications. Apart from that are 39% of the tenants of 60 years or older still in employment. This might be another indication for comparatively higher-qualified positions	Comparably well educated[44,51]	All groups of society[47]
Frequency of participation	~7 times (days) yr⁻¹	On average 2.6 times a week with big variations[44]	From once a year to more than once a week[46,47]
Know-how and skills	Primarily formal instructions in the morning and professional supervision during the whole day	Individual learning by doing, learning from neighboring tenants[44,51]	Collective restoration and stewardship activities supervised by more experienced volunteers and professionals; individual monitoring activities[45-47,49]
Innovation in cultivation	No intention for innovation in cultivation, methods or varieties	Trying something new, such as exotic plants, new cultivation methods, organic production[44]	n.a.
Property rights	Retired farmers lease out nearly all of their rice terraces, the rest is cultivated by themselves or falls abandoned; 90% of fees stay within the PA, only 10% goes to the landowners	Active farmers directly lease out small parts of their arable land to tenants, the rest is cultivated by themselves; 100% of fees goes to farmers; farmers' profit can be between 12 or 39 € per working hour (for 30 or 39 plots)[44,51]	On private and state land, sometimes use agreements with land owners[47]
Infrastructure	Big club house with showers and toilets	Children's playground, benches, toilets[44,51]	n.a.

Urban people as paddy farmers   337

ambition to plant exotic/rare plants which the neighboring plots do not have[44]. In contrast, the Japanese tenants meet for around seven collective working days a year for synchronized planting, weeding and harvesting the same variety of rice.

All of the self-harvesting farms just offer a small share of their land for rent, whereas the retired Japanese landowners lease out nearly everything. The continuity of (the rice) cultivation and the conservation of the landscape scenery is their main motivation. Austrian farmers contrarily are also motivated by economic advantages. An agricultural journal targeted to farmers in Austria and Germany[51] promotes self-harvesting with hourly revenue rates from 12 to 39 € per working hour invested by the farm family (100% of the rent goes directly to the farmers). In Japan, only 10% of the tenant's rent goes to the farmers, the rest to the PA. Moritz[51] also points out the advantage for farmers of the self-harvesting system of having the rent in their bank account before harvest time and without any weather risk. Direct marketing of farm products to self-harvesters also contributes to the farm income[44].

Neither Japanese nor Austrian tenants are motivated by producing cheap food. It is rather about producing something you cannot buy at supermarkets: self-cultivated and harvested food. The self-harvest initiative can be seen as an organic local food system in reaction to the Chernobyl disaster[51]. The Ownership System, however, is based on conventional food and might be a consequence of false labeling scandals in the 1980s and 1990s and a general mistrust in the globalized food system in Japan. In the mid-1990s—parallel to the Ownership System—a local food movement called chisan–chisho (literally translated as 'locally produced, locally consumed') appeared and sprouted all over Japan[59,60]. Local food is perceived by the Japanese as safer, more delicious, trustworthy, environmentally friendly and as a means of boosting the local economy[61].

Ownership System versus conservation volunteering

Mühlmann et al.[47] and Bell et al.[45] categorized different types of voluntary conservation organizations:
- Local associations supporting the stewardship of their local landscapes as their place of living and recreation.
- Predominantly urban voluntary tourists engaging in restoration projects or on organic farms, i.e., for landscapes outside of their living environment.
- Corporate volunteering activities.

An example for the first type is the 'Aktion Heugabel' in Austria, where local non-farmers help farmers to conserve the local landscape with its extensively managed grasslands[41]. The second group of organizations place volunteers on organic farms (e.g., 'WWOOF': World Wide Opportunities on Organic Farms) or into conservation projects worldwide[49]. An example for corporate volunteering is the company 'swiss.com' which organizes the so-called 'Nature-Action-Days' in cooperation with the WWF (World Wide Fund for Nature), where employees can engage in landscape stewardship, e.g., mowing wet meadows[47].

Apart from most of the local associations, the engagement tends to be short term. Work holidays or landscape stewardship days are time limited and confined to a particular area. This reflects a general trend in Western voluntary organizations: away from lifelong participation toward targeted and temporary engagement[62]; long-term membership becomes less attractive[63,64]. Local clubs, who are committed to the long-term cultivation of endangered landscapes (such as wet meadows) are based on volunteers, motivated to take care for the landscape they live in. But they also have difficulties in finding young members, willing to engage for a longer period[49]. This development in conservation volunteering is in contrast with the long-term urban–rural coalition of the Japanese Ownership System.

In contrast to the Ownership System, wherein the production of own rice is an important element, food production does not play any role in the voluntary conservation activities.

Social experience, however, and the management of relations between experts and professionals on the one hand and amateurs on the other hand[45] are motivating factors in both schemes.

Conclusions

The Japanese population highly values traditional cultural landscapes for tradition, culture, identity, biodiversity, quality of life and recreation. In the context of unwanted landscape changes caused by agricultural land abandonment, urban dwellers take over the responsibility for landscape stewardship. They invest time, manual labor and money to support the conservation of the traditional cultural landscape.

In contrast to self-harvesting initiatives in Europe concentrating on organic food production or volunteering focusing on nature conservation, the Japanese Ownership System combines:
- the volunteers' efforts for the common 'good' landscape;
- the individuals benefit by producing their own local food.

The success of the Ownership System might be explained by the fact that motivations for common purposes such as landscape conservation work are supported by individual benefits, such as rice or social benefits.

The urban tenants of the highly urbanized Japanese society appreciate the opportunity to accompany all steps of food production and its consumption, the sensual perception of nature, seasons, weather, phases of growth and decay, new experiences or positive effects on the education of their children. The practical rice cultivation is also, for many, a way to come closer to culture and tradition. The long-term rural–urban cooperation does not only yield satisfaction to the urban volunteers, it is highly beneficial for the conservation of agro-biodiversity and the cultural landscapes. And last but not least, the local landowners are valued and can hand on their experience and skills and thus still contribute to the cultivation of their land despite their retirement age.

A comparison with the European self-harvesting initiatives and conservation volunteering in North America, Australia, North and Central Europe demonstrates that the Ownership System is special because of its collective nature and long-term institutionalized involvement (see Table 1). However, all those forms of voluntary work on farm land reflect a clear shift from food production in the postwar period to non-productive motivations such as nature conservation, education and leisure.

In Japan, the widely spread Ownership Systems are perceived as a promising way of stopping or slowing down the process of land abandonment. However, it remains unclear whether it will help Japan to engage non-farmers to take up farming as a profession, which is also one of the objectives of the Ownership System. It might be one of a bundle of measures to keep up agricultural land use also in the less favored mountainous areas.

Technically, the Ownership System can be transferred to any kind of traditional land use that is characterized by manual labor and the production of some commodity, as for example, wine in wine terraces, or fruits in traditional orchards. The 'conserving' areas, however, should be close to urban agglomerations. Given that the prospective tenants should be able to produce and use their own products, one of the most threatened elements of rural landscapes in Austria—low-productivity grasslands, which are endangered by abandonment and/or afforestation—are not suitable for the Ownership System, as grass and weeds cannot be eaten. From the social-organizational point of view, the long-term urban–rural collaboration poses a challenge for transferability. This stands in clear contrast to tendencies in Europe, Australia and North America where citizens are reluctant toward a long-term commitment to an organization[61]. They tend to favor project-based short-term activities, combining the need to do something useful with the need for recreation and socializing.

Acknowledgements. This paper is an outcome of a PhD project. The research in Japan was funded by the Ministry of Education, Culture, Sports, Science and Technology of Japan (MEXT) and the German Academic Exchange Service (DAAD). The authors would like to express their deep gratitude to the Ōyama-sanmaida Preservation Association for providing information and supporting this survey, especially to Mitsuji Ishida, Yoshiko Sudo, Hitomi Taira and to all participants of the Ownership System, which agreed to participate in the questionnaire poll. We are beholden to Drs Nobuhiko Sawai, Kentaro Aoki and Ayako Toko, who assisted in the translation process of the questionnaires. Sincere thank also to Professor Dr Wolfgang Holzner, PhD supervisor of Pia Kieninger, for encouraging the studies for this paper. Thanks to the colleagues of the Institute of Nature Conservation Research (BOKU University) as well as to the anonymous reviewers and the editors of RAFS, for their valuable annotations to this paper.

References

1 Moore-Colyer, R.J. 2004. Kids in the corn: School harvest camps and farm labour supply in England, 1940–1950. Agricultural History Review 52:183–206.
2 Moore-Colyer, R.J. 2006. The call to the land: British and European adult voluntary farm labour; 1939–1949. Rural History 17:83–101.
3 Stenseke, M. 2009. Local participation in cultural landscape maintenance: lessons from Sweden. Land Use Policy 26: 214–223.
4 Swanwick, C. 2009. Society's attitudes to and preferences for land and landscape. Land Use Policy 26:62–75.
5 Lyson, T.A. 2004. Civic Agriculture: Reconnecting Farm, Food, and Community. University Press of New England, Lebanon, NH.
6 Allen, P. and Guthman, J. 2006. From 'old school' to 'farm-to-school': neoliberalization from the ground up. Agriculture and Human Values 23:401–415.
7 Bagdonis, J.M., Hinrichs, C.C., and Schafft, K.A. 2009. The emergence and framing of farm-to-school initiatives: civic engagement, health and local agriculture. Agriculture and Human Values 26:107–119.
8 Saldivar-Tanaka, L. and Krasny, M.E. 2004. Culturing community development, neighborhood open space, and civic agriculture: the case of Latino community gardens in New York City. Agriculture and Human Values 21:399–412.
9 DeLind, L.B. 2002. Place, work, and civic agriculture: common fields for cultivation. Agriculture and Human Values 19:217–224.
10 Takeuchi, K., Brown, R.D., Washitani, I., Tsunekawa, A., and Yokohari, M. (eds.) 2003. Satoyama – The Traditional Rural Landscape of Japan. Springer-Verlag, Tokyo, Japan.
11 MAFF – Ministry of Agriculture, Forestry and Fisheries, Japan 2009. Annual Report on Food, Agriculture and Rural and Rural Areas in Japan FY 2009 (Summary). Available at Web site http://www.maff.go.jp/e/annual_report/2009/pdf/e_all.pdf (verified February 8, 2011).
12 MAFF – Ministry of Agriculture, Forestry and Fisheries, Japan 2008. Annual Report on Food, Agriculture and Rural Areas in Japan FY 2008. Policies on Food, Agriculture and Rural Areas in Japan FY2007. Summary (Provisional Translation). Available at Web site http://www.maff.go.jp/e/annual_report/2008/pdf/e_all.pdf (verified June 22, 2010).
13 Kobori, H. and Primack, R.B. 2003. Conservation for Satoyama, the traditional landscape of Japan. Arnoldia 62:2–10.
14 Fukuda, H., Dyck, J., and Stout, J. 2003. Rice Sector Policies in Japan – Electronic Outlook Report from the Economic Research Service. Economic Research Service, USDA, p. 1–19. Available at Web site http://www.ers.usda.gov/publications/rcs/mar03/rcs030301/rcs0303-01.pdf (verified February 17, 2011).
15 MAFF – Ministry of Agriculture, Forestry and Fisheries, Japan 2003. Fact Sheet No. 1. Available at Web site http://www.maff.go.jp/e/pdf/factsheet.pdf (verified June 22, 2010).
16 Takeuchi, K. 2001. Nature conservation strategies for the 'Satoyama' and 'Satochi', habitats for secondary nature in Japan. Global Environmental Research 5:193–198.
17 MOE – Ministry of the Environment, Nature Conservation Bureau (ed.) 2010. Biodiversity is Life. Biodiversity is our Life. The National Biodiversity Strategy of Japan 2010.

18 Kuramoto, N. 2003. Citizen conservation of Satoyama landscapes. In K. Takeuchi, R.D. Brown, I. Washitani, A. Tsunekawa, and M. Yokohari (eds). Satoyama – The Traditional Rural Landscape of Japan. Springer-Verlag, Tokyo, Japan. p. 23–39.

19 Yamaji, E. 2006. Enjoyment of rural amenities by ownership program of rice terraces. Journal of Rural Planning Association 25:206–212 (in Japanese).

20 Watanabe, T. 2003. Present problems of hilly and mountainous areas under enforcement of the direct payment system. In PRIMAFF Annual Report 2003. Available at Web site http://www.maff.go.jp/primaff/koho/seika/annual/pdf/an2003-6-7.pdf (verified February 17, 2011).

21 Aono, S., Kaga, H., Shimomura, Y., and Masuda, N. 2005. Study on landscapes attractiveness to residents from the viewpoint of topographical features in agricultural area at the edge of Senboku Hill. Journal of the Japanese Institute of Landscape Architecture 68:753–756 (in Japanese, with English abstract).

22 Shibata, R. and Masuda, M. 2001. Sustainability of tanada owner system: A case study on obasute tanada, Koshoku City, Nagano prefecture. Bulletin of Agricultural and Forestry Research Center, University of Tuskuba 14:19–28 (in Japanese, with English abstract).

23 Agency for Cultural Affairs, Department of Cultural Properties, Division of Monuments and Sites, Japan (ed.) 2003. Nihon no bunkatekikeikan. Nōrinsuisangyō ni kanren suru bunkatekikeikan no hogo ni kansuru chyōsakenkyū hōkokusho (in Japanese).

24 MAFF – Ministry of Agriculture, Forestry and Fisheries, Japan 2009. Heisei 20 nendo. Chūsankanchiiki tō chokusetsushiharai seido no jisshijōkyō. Nōson shinkō kyoku. Available at Web site http://www.maff.go.jp/j/nousin/tyusan/siharai_seido/pdf/h20_zissi_data3.pdf (verified February 17, 2011) (in Japanese).

25 Takao, K., Maeda, M. and Nonami, H. 2003. Residents' perception of procedural justice in implementing the rice terrace ownership system in Asuka village, Nara prefecture. Journal of Rural Planning Association 22:26–36 (in Japanese, with English abstract).

26 Yamamoto, W., Yamaji, E., and Makiyama, M. 2002. Consciousness of rural people for ownership program of rice terraces. A case study of oyama-senmaida ownership program in Kamogawa City. Journal of Rural Planning Association 21:115–120 (in Japanese, with English abstract).

27 Yamamoto, W., Makiyama, M., and Yamaji, E. 2003. Continuity of the labor support of local farmers in 'Ownership Program' of rice terraces: case study in Ohyama District, Kamogawa City. Journal of Rural Planning Association 22:112–121 (in Japanese, with English abstract).

28 Kojima, H., Osawa, S., and Katsuno, T. 2004. Vegetation on the terraced paddy field embankment where has introduced the owner system; Case study in Ohyama Senmaida, Kamogawa City, Chiba Prefecture. Journal of Rural Planning Association 23:1–6 (in Japanese, with English abstract).

29 Kurita, H., Kimura, Y., Matsumori, K., and Osari, H. 2004. A study on the relationship between physical features of terraced rice fields landscapes and their perception. Journal of Rural Planning Association 23:85–90 (in Japanese, with English abstract).

30 Ōyamasenmaida Cultural Landscape Preservation Committee 2006. Ōyama no senmaida bunkatekikeikan hozon katsuyō keikaku. Kabushiki kaisha koa, Kamogawashi (in Japanese).

31 Kamogawa City 2004. Kamogawashi tōkeishyo – heisei 16 nenban (in Japanese).

32 Agency for Cultural Affairs, Japan, Cultural Properties Department, Monuments and Sites Division, Committee on the Preservation, Development, and Utilization of Cultural Landscapes Associated with Agriculture, Forestry and Fisheries 2003. The Report of the Study on the Protection of Cultural Landscapes Associated with Agriculture, Forestry and Fisheries. Available at Web site http://www.bunka.go.jp/english/pdf/nourinsuisan.pdf (verified February 9, 2011).

33 Yin, R.K. 2002. Case Study Research: Design and Methods. 3rd ed. Applied Social Research Methods Series, Volume 5. Sage Publications, Thousand Oaks, California, USA.

34 Flick, U. 2009. An Introduction to Qualitative Research. 4th ed. Sage Publications Ltd., London, UK.

35 National Federation of Land Improvement Association 2010. Zenkoku tanada ōnā seido ichiran. Kamogawashi tanada nōgyō chiku ōnā seido. Available at Web site http://www.inakajin.or.jp/kikin/tanada/tanada_075.html (verified February 17, 2011) (in Japanese).

36 ŌSM – Ōyamasenmaida Preservation Association (NPO) 2009. Ango tsūshin Web ban Ōyamasenmaida. Home shiryō daunrōdo. Available at Web site http://www.senmaida.com/down_load/index.php (verified August 15, 2009) (in Japanese).

37 ŌSM – Ōyamasenmaida Preservation Association (NPO) 2011. Ango tsūshin Web ban Ōyamasenmaida. Available at Web site http://www.senmaida.com/index.php (verified February 17, 2011) (in Japanese).

38 Ishikawa, S., Ae, N., and Yano, M. 2005. Chromosomal regions with quantitative trait loci controlling cadmium concentration in brown rice (Oryza sativa). New Phytologist 168:345–350.

39 Nakagahra, M., Okuno, K., and Vaughan, D. 1997. Rice genetic resources: history, conservation, investigative characterization and use in Japan. Plant Molecular Biology 35: 69–77.

40 Bartholomew Ltd (ed.) 1999. The Times Comprehensive Atlas of the World – 2000 Millennium Edition, 10th ed. Times Books, London.

41 MIC – Ministry of Internal Affairs and Communications of Japan, Statistics Bureau, Director-General for Policy Planning (Statistical Standards) & Statistical Research and Training Institute 2005. Population Census 2005. Chapter II: Population by Sex and Age. Available at Web Site http://www.stat.go.jp/english/data/kokusei/2005/poj/pdf/2005ch02.pdf (verified February 17, 2011).

42 MIC – Ministry of Internal Affairs and Communications of Japan, Statistics Bureau, Director-General for Policy Planning (Statistical Standards) & Statistical Research and Training Institute 2006. Annual Report on the Labour Force Survey 2006. Available at Web Site http://www.stat.go.jp/english/data/roudou/report/2006/ft/index.htm (verified February 17, 2011).

43 MIC – Ministry of Internal Affairs and Communications of Japan, Statistics Bureau, Director-General for Policy Planning (Statistical Standards) & Statistical Research and Training Institute 2009. The Statistical Handbook of Japan 2010. Chapter 3 Economy. Available at Web site http://www.stat.go.jp/english/data/handbook/pdf/c03cont.pdf (verified February 17, 2011).

44 Vogl, C.R., Axmann, P. and Vogl-Lukasser, B. 2003. Urban organic farming in Austria with the concept of selbsternte ('self-harvest'): An agronomic and socio-economic analysis. Renewable Agriculture and Food Systems 19:67–79.

45 Bell, S., Marzano, M., Cent, J., Kobierska, H., Podjed, D., Vandinzskaite, D., Reinert, H., Armaitiene, A., Grodińska-Jurczak, M., and Muršič, R. 2008. What counts? Volunteers and their organisations in the recording and monitoring of biodiversity. Biodiversity Conservation 17:3443–3454.

46 Miles, I., Sullivan, W.C., and Kuo, F.E. 1998. Ecological restoration volunteers: the benefits of participation. Urban Ecosystems 2:27–41.

47 Mühlmann, P. 2009. Zivilgesellschaftliches engagement in der landschaft. Das Modell freiwilliger Arbeit in der Landschaftspflege. Doctoral thesis, University of Natural Resources and Life Sciences, Vienna, Austria (in German, with English abstract).

48 Coghlan, D. 2007. Insider action research: opportunities and challenges. Management Research News 30:335–343.

49 Kieninger, P. and Penker, M. 2009. Ehrenamtliches Engagement für die Kulturlandschaft. Zoll+ Österreichische Schriftenreihe für Landschaft und Freiraum 14:92–94 (in German, with English abstract).

50 Bruyere, B. and Rappe, S. 2007. Identifying the motivations of environmental volunteers. Journal of Environmental Planning and Management 50:503–516.

51 Moritz, H. 2003. Gemüsezellen an Städter verpachten. Top-Agrar Österreich, Das Magazin für moderne Landwirtschaft 4:40–42 (in German).

52 Ryan, R.L., Kaplan, R., and Grese, R.E. 2001. Predicting volunteer commitment in environmental stewardship programmes. Journal of Environmental Planning and Management 44:629–648.

53 Measham, T.G. and Barnett, G. 2008. Environmental volunteering: motivations, modes and outcomes. Australian Geographer 39:537–552.

54 Warburton, J. and Gooch, M. 2007. Stewardship volunteering by older Australians: The generative response. Local Environment 12:43–55.

55 Bruno, R. 2010. Selbsternte, Angebot, Standorte, Wien-Hietzing, Roter Berg. Available at Web site http://www.selbsternte.at/index.php?id=94 (verified June 22, 2010) (in German).

56 Marsden, T. 2003. The Condition of Rural Sustainability. Royal Van Gorcum, Assen, The Netherlands.

57 Mather, A.S., Hill, G., and Nijnik, M. 2006. Post-productivism and rural land use: cul de sac or challenge for theorization? Journal of Rural Studies 22:441–455.

58 Enengel, B. 2009. Partizipative landschaftssteuerung. Kosten-Nutzen-Risiken-Relationen aus Sicht der Beteiligten. Doctoral thesis, University of Natural Resources and Life Sciences, Vienna, Austria (in German, with English abstract).

59 Nishiyama, M. 2010. Alternative agro-food movement in contemporary Japan. In M. Tsutsumi (ed.). Turning Point of Women Families and Agriculture in Rural Japan. Gakubunsha, Tokyo, Japan. p. 281–297.

60 Yoshino, K., Katayama, C., and Morofuji, K. 2010. The present situation of local supply and consumption of agricultural products from the aspect of acquisition and utilization by local people. In M. Tsutsumi (ed.). Turning Point of Women, Families and Agriculture in Rural Japan. Gakubunsha, Tokyo, Japan. p. 261–280.

61 Kimura, A.H. and Nishiyama, M. 2008. The chisan-chisho movement: Japanese local food movement and its challenges. Agriculture and Human Values 25:49–64.

62 Schlute, R. 2006. Freiwillige in naturschutzverbänden. In S. Bremer, K.-H. Erdmann and T. Hopf (eds) Freiwilligenarbeit im Naturschutz. Naturschutz und Biologische Vielfalt, Vol. 37. Bundesamt für Naturschutz (BfN), Bonn, Germany. p. 79–90 (in German).

63 Heinze, R.G. and Olk, T. 1999. Vom ehrenamt zum bürgerschaftlichen engagement. Trends des begrifflichen und gesellschaftlichen strukturwandels. In E. Kistler, H.-H. Noll and E. Priller (eds). Perspektiven gesellschaftlichen Zusammenhalts. Empirische Befunde, Praxiserfahrung, Messkonzepte. Edition Sigma, Berlin, Germany. p. 77–100 (in German).

64 Lossing, B. and Toennes, A. 1989. Lokale bürgerinitiativen im Umweltschutz – eine schriftliche Befragung. In J. Noeke (ed.). Bürgerinitiativen und Umweltschutz. Freiwillige und ehrenamtliche Arbeit im Umweltschutz. IFU-Werkstattreihe, Vol. 18. Institut für Umweltschutz der Universität Dortmund, Germany. p. 16–37 (in German).

Artikel 3: Kieninger, P.R., Penker, M., Yamaji, E. 2012. Esthetic and spiritual values motivating collective action for the conservation of traditional rural landscapes – A case study of rice terraces in Japan. Renewable Agriculture and Food Systems DOI: http://dx.doi.org/10.1017/S1742170512000269: 1–16.

Rahmenschrift Dissertation Pia R. Kieninger

Esthetic and spiritual values motivating collective action for the conservation of cultural landscape—A case study of rice terraces in Japan

Pia R. Kieninger[1,*], Marianne Penker[2], and Eiji Yamaji[3]

[1]Institute of Integrative Nature Conservation Research, Department of Integrative Biology and Biodiversity Research, University of Natural Resources and Life Sciences, Vienna, Gregor Mendel Strasse 33, 1180 Vienna, Austria.
[2]Institute of Sustainable Economic Development, Department of Economics and Social Sciences, University of Natural Resources and Life Sciences, Vienna, Feistmantelstrasse 4, 1180 Vienna, Austria.
[3]Department of International Studies, Division of Environmental Studies, Graduate School of Frontier Sciences, The University of Tokyo, 5-1-5 Kashiwanoha, 277-8653 Kashiwa City, Japan.
*Corresponding author: kieninger.pia@gmail.com

Accepted 17 July 2012   Research Paper

Abstract

Japan's rice terrace landscapes are not only used for food production but also appreciated as a place of high biocultural value. This paper pursues the question as to how far esthetic and spiritual values influence the motivation to participate in collective agricultural actions aiming at the conservation of traditional land use systems, the respective cultural (= traditional rural) landscapes and their biocultural diversity. Our results show that in the Ownership System of Ōyamasenmaida (Chiba Prefecture, Japan) landscape beauty is the main motivator for the mainly urban volunteers (the 'tenants') to participate in activities of the local Preservation Association, as well as for visitors who merely come to enjoy the scenery of the rice terraces. The active tenants, however, differ from the 'passive' visitors in their ecological interest and emotional attachment to the area. Interestingly, there is also a difference regarding the belief in nature spirits. A higher percentage of people who can imagine that such spirits are always present have been found among the tenants than among the visitors. Even more significant in this respect was the difference between female tenants and female visitors. To what extent spirituality is the cause for or an effect of involvement in nature conservation activities cannot be concluded from this survey. Future studies should therefore take a closer look at the connection between spirituality/religiosity and engagement in nature conservation activities. In Western countries (mainly in Central Europe or North America), nature conservation works on a more 'scientific' level, mobilizing engagement through scientific evidence on, for example, losses of species or biodiversity. Addressing the motivations of the volunteers on an emotional, esthetic or social level could be a promising way forward.

Key words: esthetics, biocultural diversity, collective action, conservation of traditional rural landscape, motivation, Ōyamasenmaida (Chiba Prefecture, Japan), rice terraces, *satoyama*, spirituality, *Tanada* Ownership System

Introduction

Aims of this article

Traditional land use systems and the respective cultural (= traditional rural) landscapes of high biocultural diversity are under great pressure due to technological and socio-economic changes and in many parts of the world are not economically viable anymore. If the market fails, some kind of action is needed to preserve this biocultural diversity. Thus, mobilization and motivation of volunteers to participate in the management of traditional land use systems become important factors. During field research into the biodiversity of a rural paddy landscape in Central Japan (see subsection 'Study site') the first author became aware that hard scientific facts, such as the conservation of species, were not enough to motivate

© Cambridge University Press 2012

people to engage in such activities. Esthetic and spiritual values might be even stronger motivators. In this article, we compare a group of mainly urban volunteers who invest personal time, money and manual labor to contribute to the conservation of a terraced rice landscape in Japan with a group of visitors 'passively' admiring the same terraced landscape. Both groups might appreciate the beauty of the landscape; however, those actively contributing to conservation and agricultural management might be driven by additional motivations. Knowledge about the factors triggering readiness to participate actively in activities can be helpful in developing effective conservation strategies.

Satoyama and biocultural diversity

As in many parts of the world, most of the Japanese landscape has been shaped by man. In the hilly and mountainous regions, which cover more than 70% of the total land area[1], small-scale agriculture, adapted to the topographic situation, formed the 1000-year-old cultural 'satoyama' landscape[2,3], consisting of villages at the foot of the mountains, paddies and other arable crop fields mainly on terraces, secondary grasslands, and coppiced forests on the slopes, conifer plantations, bamboo groves, water reservoirs, ponds, streams, shrines, temples and graveyards. For the Japanese, satoyama is often the symbol of the last remnants of a natural environment[4]. The FAO has included satoyama sites in its list of Globally Important Agricultural Heritage Systems, which are defined as 'Remarkable land use systems and landscapes which are rich in globally significant biological diversity evolving from the co-adaptation of a community with its environment and its needs and aspirations for sustainable development'[5]. Paddy fields play a key role. In total 54% of Japan's farmlands are paddy fields[6]; of these, 14.9% are terraced rice fields ('tanada' in Japanese)[7]. Apart from food production and their relevance for other ecosystem services, e.g., prevention of floods and erosion, water purification or ground water protection[1,8], paddies are important biotopes for a rich and specialized fauna and flora. A recent study, presented at the 10th UN Convention of Biodiversity, lists 5668 animal and 2075 plant species that are connected to the rice field ecosystem in Japan[9]. Also, from a cultural and religious perspective, rice played and continues to play a prominent part in Japanese daily life[10–13]. Lately, an increased awareness can be found in Japanese society regarding rice terrace landscapes, also called 'senmaida' (1000 rice fields), which are perceived by many people as the 'nature' most close to them[14], a place to which they feel a strong attachment. Tanada is the landscape of their ancestors, representing culture, tradition, (spiritual) homeland and national identity—all in all, a landscape loaded with emotions. Besides that, it is considered to be attractive due to its esthetic value[14].

Satoyama and biocultural diversity at risk

Currently, 40% of the land area is estimated in the Fourth National Biodiversity Strategy of Japan to be satoyama[2]. Its continued existence, however, is at a great risk, due to serious demographic and economic upheavals. Declining birth rates, an aging society (nearly one-quarter of the Japanese are 65 years or older[15]), small average farm size[16], high costs of farm labor[17], and thus high production costs[18] in combination with falling rice consumption[19], result in a decrease of agricultural income[1], and consequently in the migration of the younger population to the cities. This, in turn, leads to 'retirement farming' (70% of Japanese agriculture rests on the shoulders of retired people[20]) and finally to the abandonment of agriculture and farmland and to depopulation of the rural areas[1,21].

Rise of collective action

Initiatives for a revival of rural areas started in the late 1980s–early 1990s in the form of local citizens' movements as well as in government programs, and are increasing continuously[20,22–24]. Thus, the Ministry of Agriculture, Forestry and Fisheries of Japan registered joint projects between farmers and non-farmers involving 19,514 organizations nationwide operating over an area of 1.43 million ha, i.e. 31% of the Japanese cultivated land area[1,25]. Several activities introduced by the Japanese government are targeted specifically at the conservation of the tanada landscape, such as, the yearly tanada summit to strengthen the local economy[14,26], the nationwide award of the country's 100 most beautiful rice terraces[27] in order to increase their popularity and to bring citizens to engage in their protection[27], or the support of the legal implementation of the Tanada Ownership System[28]. The Tanada Ownership System is a land tenure system, where non-farmer volunteers ('the tenants'), often city dwellers, rent a piece of land (mostly rice terraces, but also other arable land) against a certain rental fee and cultivate it under the guidance of the landowners or other experienced supervisors (mainly locals)[14,19,28].

The link between esthetics and spirituality— the Japanese concept of nature

The esthetic appreciation of nature has a long tradition in Japan[29] and often also has a religious meaning[30,31]. This becomes evident in the Shintoistic deification of impressive natural phenomena in literature and poetry, in the visual arts or in the garden design of monasteries, where idealized and refined images of nature are reproduced in miniature format, sometimes reaching a maximum of abstraction and reduction and arranged in a scenic composition. In Japanese 'religious understanding'— resulting from mythology, the '"Buddha-nature" in all things'[32] and Shintoism with its proverbial myriad kami-sama—all creatures in nature have a soul, even natural phenomena and inorganic objects (*kami-sama* are divine powers, which are inherent in all things that give the onlooker a feeling of awe, such as the sun, moon, rocks, rivers, old trees, caves, flowers, animals, and even people with an outstanding personality)[12,30,32]. Moreover, mythical creatures also animate nature. Man and nature are perceived as a union, inseparable from the spiritual world[30,32]. This could be one of the reasons why Japanese people do not make the same sharp distinction between 'man-made' and 'natural' as Westerners do[32–35]. This attitude, however, does not automatically imply that Japanese people have a more heedful and harmonious relationship with nature (which is a common stereotype), but merely indicates that they appreciate the formative human hand in nature. Japanese have a preference for semi-natural rural sceneries[36,37], while many Westerners long for untouched primary nature[29,31,35–42].

At this point it must be noted that Japanese religious understanding should not be viewed from a Western point of view. 'The word, religion, is historically and culturally constructed'[43]. A large percentage of Japanese people define themselves as 'unreligious', although they regularly conduct religious rituals, festivals or other religious events[43–46]. To the Japanese 'religion' means to be a member of a religious institution, but it does not refer to religious feelings or values[43–45]. Furthermore, the Japanese do not make a sharp distinction between faith and superstition (the latter of which for them has no derogatory connotations) as Westerners do (Dr Bernhard Scheid, Department of Japanese Studies, University Vienna, Institute for the Cultural and Intellectual History of Asia, personal communication, December 16, 2011). Thus, it did not present a problem to ask a question concerning 'the belief in spirits in nature' for the purpose of this study (see subsection 'Sense of spirits in nature').

State-of-the-art of esthetics and spirituality as motivators for collective conservation activities

Studies on examples of collective conservation activities related to rural landscapes in Japan are rare[17,47–49]. Publications on the (*Tanada*) Ownership System pertain to its general structure, sustainability and procedural implementation and its role in conservation, but are available almost solely to the Japanese-speaking scientific community[26,28,50–55]. Several publications can be found dealing with the Japanese concept of nature and landscape preferences of different land users[29,31,35–39,41,42,56–64], and with the link between nature, esthetics and religion[12,13,29,31,39,65]. To our knowledge no international literature that discusses the connection between the esthetic and spiritual dimension and the motivation of people participating in collective activities for the conservation of traditional land use systems exists to date. This article follows a previous paper, where the general profile of the 'Ōyamasenmaida Ownership System', the major motivation of the volunteers and the transferability of this Japanese type of voluntary farm work activity to Western countries have already been discussed[28]. In this article, we focus on the role of esthetic appreciation and attitude toward spirits in nature and compare tenants and visitors (tourists just staying for a relatively limited time period). By means of an integrative research approach, we wish to provide not only an original contribution to science but also recommendations for the practical mobilization of volunteers for collective action aiming at the conservation of rural landscapes and their biodiversity. The volunteers in the Ownership System in Ōyamasenmaida call themselves '*ōnā*', derived from the English word 'owner'. As the Ōyamasenmaida Ownership System is a land tenure system, we have decided for the sake of clarity to use the term 'tenant' instead of '*ōnā*'.

Research Design, Methods, Data and Study Site

Research design, research questions and hypotheses

The research design (see Fig. 1) shows the overall research question which encompasses four questions:
(1) What is the motivation of tenants to engage in the Ownership System and of the visitors to see Ōyamasenmaida?
(2) What do both groups regard as the most important landscape attributes in Ōyamasenmaida?
(3) How do they perceive the landscape?
(4) What is the attitude toward nature spirits among tenants and visitors?

These questions will be answered by a comparison of the 'active' (tenants) and 'passive' (visitors) land users in Ōyamasenmaida. Their motivations and the underlying values will be surveyed based on two similar questionnaires (including photo-based landscape assessments for the tenants). We have four hypotheses:
(1) The attractiveness of the rice terrace landscape scenery plays a key role in the readiness of the tenants to engage in conservation activities and of the visitors to see the site.
(2) The attractiveness of the rice terrace landscape scenery is an important landscape attribute for the tenants and visitors.
(3) The traditional rural landscape (*satoyama*) is often perceived as (untouched) nature.
(4) Visitors and tenants imagine differently how this 'nature' is inhabited by spirits.

A case study approach was determined for this study, as the research is of an exploratory nature and addresses a contemporary phenomenon within its real-life context without clear borders between phenomenon and context[66].

Connection between esthetic and spiritual values and voluntary action for biodiversity conservation

(1) What is the motivation of the tenants to engage in the Ownership System and of the visitors to come to see Ōyamasenmaida?
(2) What do both groups regard as the most important landscape attributes in Ōyamasenmaida?
(3) How do they perceive the landscape?
(4) What is the attitude towards nature spirits among tenants and visitors?

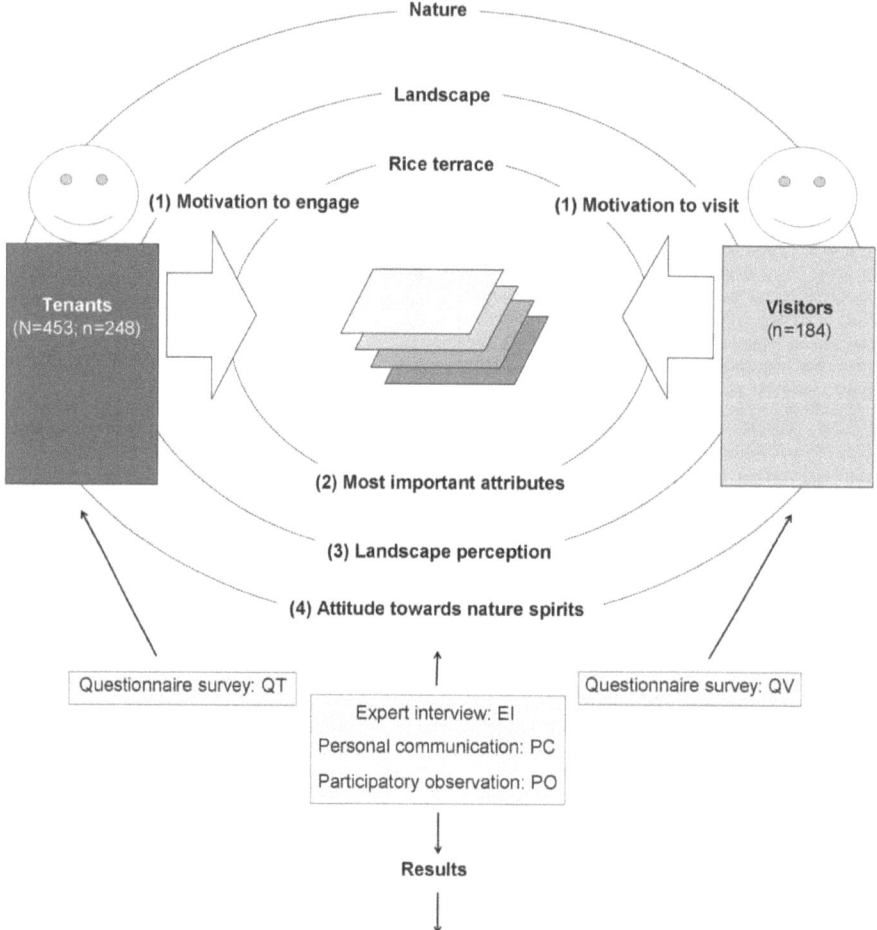

Figure 1. Research design.

Data collection

For this study, in total more than 50 days (December 2004–May 2006) were spent by the corresponding author on the site, taking part in various general and tenant meetings, tenants' working days (e.g., rice planting, weeding and harvesting), traditional cultural events, festivals and everyday life. It was by no means simple to become acquainted with the local people on site. As they were not accustomed to a foreign woman researcher, not only did general communication difficulties present an obstacle, but it also took nearly 1 year until they trusted her and lost their reservations to talk. After this warming-up phase, however, the 'Ōyamasenmaida people', particularly those from the Preservation Association, supported the study wholeheartedly.

The focus of this article lies in the information collected with structured questionnaires. The questionnaire survey is a cross-sectional comparison of two groups at a single point in time. Additional qualitative data are the results of the expert interview (EI), personal communication (PC), and participatory observation (PO).

Based on PC, PO and first explorative interviews with tenants and representatives of the local Ōyamasenmaida Preservation Association (PA), two questionnaires were designed in English for the two groups of 'tenants' (QT) and 'visitors' (QV), discussed with colleagues from the University of Tōkyō and then translated into Japanese. As e-mail addresses were mostly unknown, the questionnaires were distributed and collected directly on site (visitors and tenants at meetings) or sent by post, together with a stamped envelope for reply. Names and postal addresses of the tenants were provided by the PA. In a cover letter, the head of the PA motivated the tenants to participate in the study.

The questionnaires include questions on attitude, behavior and classification in an open, semi-closed and closed style. The open answers were transcribed and translated from Japanese into English. Direct quotations by the tenants or visitors recorded in the questionnaires or the statements from the EI with the director of the PA are therefore, in most cases, translations from Japanese into English. The translation from Japanese into English (particularly of the open statements in the questionnaires) was at times difficult, as Japanese is a very context-dependent language and the meaning of single short sentences is not always easy to grasp.

Study site

The case study looks into the (*Tanada*) Ownership System of Ōyamasenmaida, a traditional rice terrace landscape covering an area of 3.1 ha[67] in the midst of the hills on the Bōsō peninsula, around 15 km west of Kamogawa City and 100 km south-east of Tōkyō. In 1999, the rice terraces of Ōyamasenmaida were designated by the government as one of the 'Top 100 Terraced Paddy Fields of Japan'[27].

Before this designation, the landowners were already aware of the particularity of their rice terraces and in 1997, together with the locals, founded the NPO Ōyamasenmaida PA (PC: director, June 18, 2005). In order to secure the future management—eight of the 11 proprietors were already at retirement age—the PA drew up an Ownership System, which started officially in 2000 (PC: director, June 18, 2005).

The Ownership System of Ōyamasenmaida offers five different programs for rice, soybean, cotton/indigo and rice wine, respectively, and also organizes traditional dancing, cooking, and handicraft events and nature experience 'classes' (e.g., firefly watching) open to everyone[28,68].

Through its closeness to Tōkyō as well as through the popularity of Ōyamasenmaida, nurtured not least through documentaries and advertising campaigns (PO: April 2005 and November 2010; PC: informant A and B September and November 2010), the run on it is enormous (EI: director, June 18, 2005). Even the Emperor of Japan has already been there[69]. This is why the Ownership System gradually increased to more than 4.5 ha[67], partly adjacent to the main area, partly also on other sites around 1 km away (unpublished data; PO: October 16, 2005).

Principal Findings

In this section, we seek to compare the motivation of the 248 tenants and 184 visitors to engage in the Ownership System or to visit Ōyamasenmaida, and their underlying esthetic and spiritual values.

Motivations

The conservation of the traditional rural landscapes is of great importance for visitors and tenants alike. Of the 241 tenants, 238 (98.8%) believe that traditional rural landscapes such as Ōyamasenmaida have to be maintained ($n=248$, n.a.=7). Agreement in favor of maintenance among the visitors lies at 97.6% (164 persons) ($n=184$, n.a.=16).

The main motivations for participation in the Ownership System (Fig. 2) are simply 'love' for the rice terraces (119 affirmations, 50.2%), 'being close to nature', and 'taking part in the PA' (105 and 102 affirmations). The conservation of the rice terraces itself lies in fourth place. From an economic point of view, the Ownership System has no real relevance as a straight food production system, since the size of the fields is relatively small and the self-produced food altogether comparatively costly—it would be cheaper to buy the rice in the supermarket. This is possibly also an explanation for why only 10 persons indicate that their participation is directly due to the production of rice, soybean or rice wine and dyeing. Even if the self-produced goods are not the key impetus for participation, it can be assumed that they are an

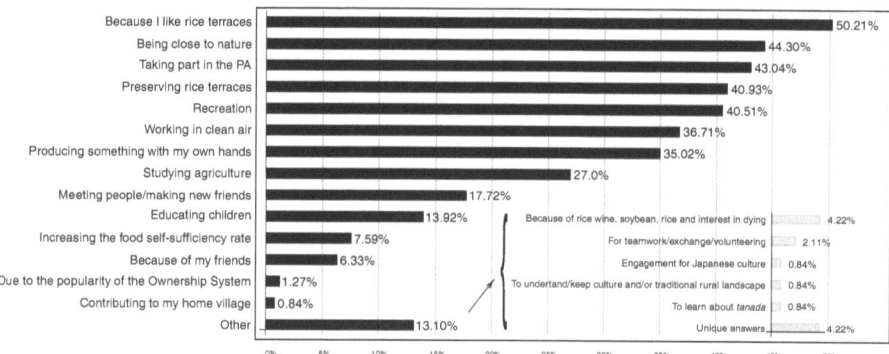

Figure 2. Tenants' motivation for participating in the ownership system (897 valid answers from 237 persons) given in average numbers (figure based on Kieninger et al.[28]). The question is semi-open (the indication of a maximum of five reasons for participating was allowed). Apart from the 14 preformulated answer categories, there was also the possibility to provide an individual answer. These open answers were coded and are shown together with the respective percentages on the right-hand side of the figure.

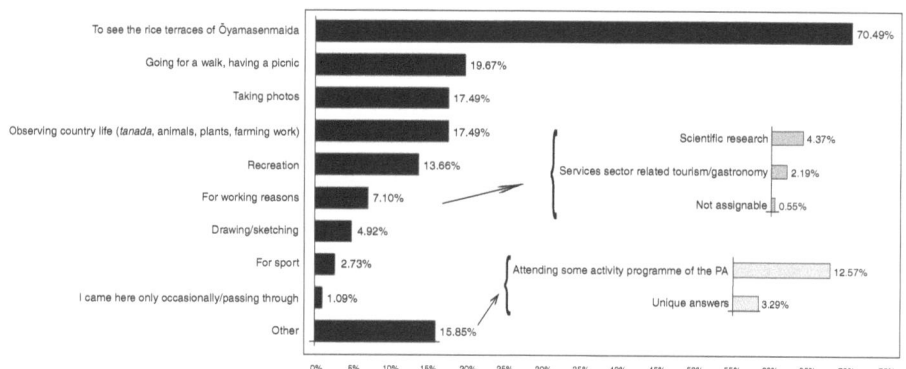

Figure 3. Visitors' motivation to come to Ōyamasenmaida (312 valid answers from 183 persons) given in average numbers. The question is semi-open (the indication of a maximum of five reasons for participating was allowed). Apart from nine preformulated answer categories, there was also the possibility to provide an individual answer in the category 'other' and 'for working reasons'. The open answers are shown together with the respective percentages on the right-hand side of the figure.

important 'by-product' with an ideal value, which the tenants are proud of: 'They eat the harvested products themselves or give them to friends on special days, e.g. on commemoration days or for their friends' party.' (EI: director, November 14, 2010).

The main reason (70.5%) why the visitors come to Ōyamasenmaida is to see the rice terraces (Fig. 3). Almost all of the other reasons have some connection with the rice terraces too. Among the open answers the attendance of traditional cultural events such as sushi-, tofu- or miso-producing festivals prevails. These 'festivals' are traditional events related to the cultivation of rice and soybeans (on some of the rice terraces soybeans are grown[28]), and the visitors come to experience them in the authentic 'environment'. In summary, 92.6% of the visitors come to Ōyamasenmaida because of the terrace landscape.

Esthetics

Attractive attributes of Ōyamasenmaida deserving protection. In order also to obtain an insight into the local attributes most attractive to visitors and tenants, they were asked to select from a list of predetermined

Table 1. Important traditional rural landscape attributes from the viewpoint of visitors ($n = 184$; 17 persons chose one attribute, 32 two and 120 three; n.a. = 15) and tenants ($n = 248$; six persons chose one attribute, 72 and 195 three; n.a. = 40). P-values given for the Pearson's chi-squared test with Yate's continuity correction; P-values for significant terms at the 0.001-level are given in bold.

	Visitors	Tenants	P-value (χ^2)
Scenic landscape beauty of the rice terraces	84.02%	63.00%	**<0.001**
Entire landscape with rice terraces, mountains, forests, etc.	63.91%	51.92%	0.025
Harmony of man and nature	30.20%	46.63%	0.002
Typical *satoyama* fauna, such as birds, frogs, salamanders, butterflies, caterpillars and fireflies	6.51%	30.29%	**<0.001**

Other arguments: 'To activate regional economy' (QT 95); 'Croaking of the frogs' (QV 54 & 92); 'Culture' (QV 19); 'That Ōyamasenmaida is in good condition' (QV 31); 'Mr. Ishida [annotation: the director] is hard working and cheerful' (QV 51); 'Rain and the rice terraces' (QV 55); 'Company, meeting people, talking, community, making friends' (QV 72); 'Miso-making experience' (QV 117); 'Rice terraces prevent landslides' (QV 170); 'The activity to vitalize this area' (QV 179).

answers what they regard as most impressive (visitors) or most important to conserve (tenants). A maximum of three answers was allowed. Visitors could choose from nine, and tenants from eight possibilities (Table 1). Additional personal comments were provided by 10 visitors and one tenant.

For both visitors and tenants the 'scenic landscape beauty, particularly the rice terraces' is the most attractive attribute [142 visitors ($n = 184$, n.a. = 15) and 131 tenants ($n = 248$, n.a. = 40); see also Table 1]. Also the 'entire landscape' (108 visitors and 108 tenants) and, thirdly, the 'harmony of man and nature' are very often mentioned, the latter, however, with a reverse weighting, as this attribute seems to be more relevant for the tenants. Clearly visible is the differing importance of the 'typical *satoyama* fauna, such as birds, frogs, salamanders, butterflies, caterpillars and fireflies' between tenants and visitors. For 63 tenants, the fauna belongs to the most important things to conserve in Ōyamasenmaida, but only 11 visitors regard it as the most impressive part of this landscape. This difference is highly significant.

Landscape perception. From Figure 4 it is clear that both groups perceive the landscape as beautiful: 223 tenants (97.81%) and 146 visitors (91.25%) describe the landscape of Ōyamasenmaida as 'beautiful' or 'very beautiful'; the tenants, however, give slightly higher, albeit significant, valuations, as they perceive 'their' landscape, the place where they work with their hands, as significantly more beautiful than the visitors just passing by. Significant differences can also be identified between the two groups regarding the 'inspiring quality' and the 'uniqueness' of the landscape. Very high agreement in numbers and on average (Fig. 4) can be detected on the landscape parameters 'harmony of man and nature' and 'traditional (landscape)'.

A very surprising result from a European point of view, and confirming our hypothesis that nature and cultural landscape (traditional rural landscape) are comprehended by Japanese people as a union, is that even though Ōyamasenmaida—a 1000-year-old rural landscape formed and shaped by humankind—is characterized by 83.78% of the tenants ($n = 248$; n.a. = 26) and 87.74% of the visitors ($n = 184$; n.a. = 29) as a traditional/very traditional landscape, it is described by a comparatively high number of tenants (27.95%; 64 of 248, n.a. = 19) and visitors (29.75%; 47 of 184, n.a. = 26) as a 'very untouched'/'untouched' landscape. The director of the Ōyamasenmaida Ownership System says this: 'The nature of Ōyamasenmaida is not the 'real nature'. People created it.... It is not a mountain or a virgin forest without human influence, but a rice field made by humankind.... In Minamiboso, Chibaken, Kamogawa there is no... forest, which one can call a virgin forest. What the tenants define as nice nature is cultural landscape.' (EI: director, November 14, 2010).

Overall both groups perceive the landscape similarly.

Photo-based landscape evaluation. A photo of an 'ordinary' contemporary Japanese paddy field (Fig. 5) was inserted in the questionnaire. The tenants were asked in a closed question (yes/no) whether they would participate in the activities of the PA if the Ōyamasenmaida landscape (Fig. 6) resembled that shown in Figure 5; in a second step they were asked to explain their decision in an open answer with 'why yes/why no'.

Of the 225 respondents ($n = 248$, n.a. 23), 82.22% would cease their participation in Ōyamasenmaida if the landscape resembled Figure 5, while 17.77% (13 female and 27 male) would continue. The open answers were translated and coded (Figs. 7 and 8). The most frequent category (36.17%) of 'Yes, I would participate because...' (Fig. 7) is: 'I would like to experience and/or support agriculture'. For these interviewees, the agricultural activity itself has priority rather than the environment or landscape, e.g., 'I think the agricultural activity would be the same' (QT 28). 'The main purpose was to experience farming activities for myself' (QT 34). 'It is still connected to agriculture' (QT 204). 'My interest is in agricultural production. If the area is large, I will participate [by cultivating] a part of it' (QT 219).

The second most frequent categories expressing acceptance were: 'This landscape is also nice and/or it is also nature' and 'My main focus is not *tanada* and/or nice scenery' (each with eight answers). These tenants regard the rice field shown in Figure 5 as a good place of nature or rather state directly that the main focus of their participation in Ōyamasenmaida is not nice scenery or

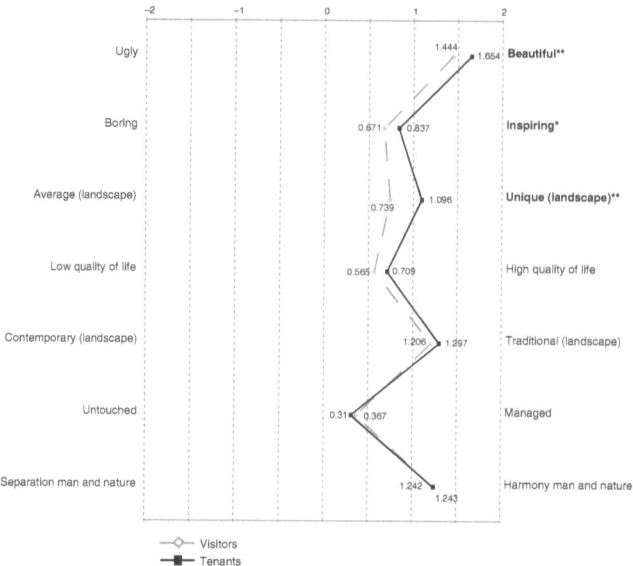

Figure 4. View of tenants and visitors regarding different landscape parameters, measured with the semantic differential on a 5-scale bar[70]. In a *t*-test, significant differences in the landscape perception of Ōyamasenmaida between visitors and tenants can be found for the parameters 'boring–inspiring', 'standard–unique (landscape)' and 'ugly–beautiful'. Significant differences between mean values are marked with *, if significant at the 0.05 level, and with **, if significant at the 0.01 level.

Figure 5. In contrast to the rice terraces of Ōyamasenmaida (Fig. 6) this paddy field (Chiba prefecture) is comparatively large. It lies in the plain and is located adjacent to a bicycle/footpath and a straight single lane road.

tanada: 'Yes, I would participate because': 'I am interested in agriculture and nature, even if it is not *tanada*.' (QT 36), 'Landscape is great.' (QT 127), 'I want to get in touch with nature.' (QT 146) or 'I would like to experience agriculture itself rather than merely enjoying the scenery' (QT 153), etc.

Figure 6. The *tanada* landscape of Ōyamasenmaida is located in the midst of the hills (80–150 m above sea level). The size of the terraces ranges from 20 to 900 sqm[67]. Forest and bamboo groves are in the background. The road skirts the terraces on the right.

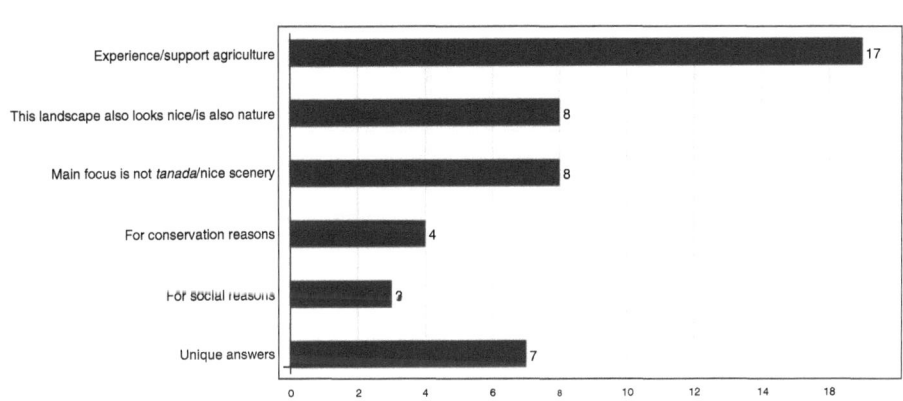

Figure 7. Reasons for continuing the ownership system on a paddy field such as the one shown in Figure 5 (47 arguments from 35 people).

Most prevalent among 'No, I would not participate because...' (Fig. 8) is with 21.47% of the category 'Common place and/or not attractive', followed by, and often linked to 'It's not a *tanada* landscape' (19.41%), e.g., 'I can see this kind of landscape anywhere. If it were not *tanada*, there would be no special reason to visit Ōyamasenmaida.' (QT 17), 'In my neighborhood there are paddies, but I am attracted by *tanada*.' (QT 71), 'Such scenery can be seen in my neighborhood. One of the main reasons why I am participating in the activity is that *tanada* has the most attractive landscape.' (QT 76).

The third most important point for the tenants is handwork (14.71%). As they want to experience agricultural work manually, large and industrialized fields such as that in Figure 5, where machines can be used for work, are of no interest to them, e.g., 'The charm of *tanada* is to experience nature without machines.' (QT 94), 'This field can be cultivated by machines and the agricultural environment can therefore be maintained.' (QT 173), 'I wanted to obtain experience in agricultural handwork in a rice field, where animals and plants live together.' (QT 200), 'This area is more suitable for machines. *Tanada* can only be cultivated by hand. It is important to protect the heritage of our ancestors and the cultural landscape. I could bring my children and grandchildren to learn to work with their hands. It is very

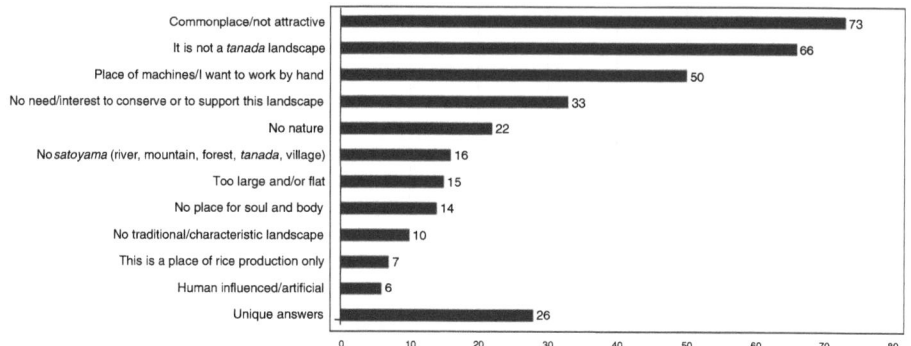

Figure 8. Reasons for ceasing participation if Ōyamasenmaida resembled the plain paddy field shown in Figure 5 (340 arguments from 168 people).

important that children see that handwork is exhausting.' (QT 211), '*Tanada* landscape is important. Working with machines is no fun. To become active in agriculture is not my goal.' (QT 221).

The evaluation of the picture clearly shows that beautiful scenery is a highly relevant motivator for the tenants. Scenic beauty is thereby tightly linked to *tanada*, cultivated in the traditional way by hand. Plain, large and mechanized fields are neither accepted nor are they considered by most of the tenants as something worth protecting or supporting: 'The *tanada* landscape seems to me the archetypal scenery of Japan. I am not attracted to places like the one on this photo, which is not *satoyama*.' (QT 163), 'Different from *tanada*, the scenery is flat and not interesting.' (QT 206), 'I didn't just want to engage in agriculture/farming but I was also very interested in the unique scenery of senmaida and its protection.' (QT 226). Also, the director of the Ownership System confirms a certain preference of the tenants for a rice terrace landscape: '... they want to participate [in the Ownership System] in a nice landscape, such as Ōyamasenmaida. Thus ... as the city dwellers themselves partake in the Ownership, as they become tenants and take part, thus, the magnificent beautiful landscape is conserved. That is joyful for the city dwellers, for the tenants' (EI: director, November 14, 2010).

Spirituality

In the last question in the questionnaires we asked tenants ($n = 248$, n.a. $= 14$) and visitors ($n = 184$, n.a. $= 18$) whether they have a certain 'feeling' for spirits in nature: 'Do you ever have the feeling that in certain parts of nature such as, for instance, in the mountains, valleys, rice terraces, orchards, streams, lakes, plants, and trees, some kind of spirits live?' We use the term 'spirituality' for the 'belief' in spirits (deities or other immaterial, mystical or mythological beings/powers) inherent in nature. It could be questioned, whether 'to feel spirits' means to believe in them (PC: Dr Brigit Staemmler, Department of Japanese Studies, University Tübingen, January 13, 2012). We use 'belief' in the sense defined by Merriam-Webster, which is: 'belief is a mental acceptance but may or may not imply certitude in the believer'[71].

Two hundred and thirty-four tenants and 166 visitors gave an answer to this question on a five-point scale from 'always', 'often', 'sometimes', 'rarely' to 'not at all'. Additionally, even if not intended, three tenants and one visitor added personal annotations e.g., 'Everything is divine.' (QT 223), 'All nature's workings are divine secret.' (QT 215), 'I believe in spirits in the mountains, valleys, terrace fields, little rivers, plants and trees, but not in lakes and orchards.' (QV 94).

More than half of the tenants (133 persons, 56.84%) have an impression of spirits in nature (18.8% always, 14.53% often, 23.5% sometimes), while 19.66% rarely and 23.5% do not believe in them at all. If we compare this result with the visitors' sample, we can see that in this group the feeling of spirits in nature (always–sometimes) is a bit lower, by 50%. A significant difference can be identified regarding 'I always have the feeling of spirits in nature' between tenants (men and women, 18.8%) and visitors (men and women, 9.64%) ($P = 0.017$) and a highly significant difference between tenant$_{women}$ (25.86%) and visitor$_{women}$ (8.43%) ($p = 0.010$).

With these results, our hypothesis of a link existing between 'spirituality' and participation in voluntary nature conservation activities such as the Ownership System can be partly confirmed. Partly, because the differences are apparent only in the category 'I always have the feeling of spirits in nature'. The reason, therefore, cannot be explained completely. The director of Ōyamasenmaida argues that the belief in the 'great nature' (spirits) is a result of their own food production: 'Not necessarily only *tanada*, but the villages and the whole rice culture (in general) have a religious meaning. In the Japanese religion the whole of nature, everything, is *kami-sama*. The sun, the stones, the trees, the earth, the water, everything is *kami-sama*. Since the idea of nature as *kami-sama* is in the hearts of the Japanese people, it is not necessarily the case that particularly rice terraces have such a connotation.... People who live in the cities (don't know) that food is influenced by nature, from nature.... People who participate in the Tanada Ownership System understand this. If you live in the city and you buy food with money, you don't realize it. But if you participate here, you understand that food is cultivated in nature while it is influenced by nature. And then they perhaps think that nature in this sense is magnificent.' (EI: director, November 14, 2010). However, a connection between duration of participation in the Ownership System and an increased belief in spirits among the tenants cannot be found in our results. Also, concerning age there is only a slight difference between tenants and visitors (52.25 years tenants versus 47.8 years visitors) in the category 'I always have the feeling of spirits in nature'.

The above-mentioned highly significant difference between tenant$_{women}$ and visitor$_{women}$ in the category 'I always have the feeling of spirits in nature' is remarkable. Regarding age, no difference can be found between these two groups (tenant$_{women}$ 47.46 years, visitor$_{women}$ 45.17 years). An age difference can therefore also be excluded as a reason here.

Discussion

'From beauty to duty'[72]

It has often been proved that esthetically appealing 'objects' such as landscapes, animals or plants create a deeper attachment and greater respect toward nature, influencing people to act in an environmentally friendly manner and to engage in nature conservation[29,31,72,73]. This is the case in many countries, not only in Japan. For example, in the USA, the designation of national parks and wilderness protection in the 19th and 20th centuries was based mainly on esthetic nature appreciation[31,72]. The esthetic quality of a certain site can even be increased by a 'label of scenic designation'[74]. This happens easily in countries with a collectivist worldview[75]. In our study, we asked about the motivation of the tenants to engage in the Ownership System and about the motivation of the visitors to see the site, as well as about the most important landscape attributes. We hypothesized that the attractiveness of the rice terrace landscape scenery played a key role and was an important landscape attribute. Both of these hypotheses could be confirmed.

As Ōyamasenmaida, after its designation to the 'Top 100 Terraced Paddy Fields of Japan', was advertised in audio-visual and print media, the number of visitors, often in organized groups, increased rapidly. With a relatively limited length of their stay, their main motivation is to visit the site in order to appreciate the terrace landscape scenery. For the tenants, the situation, however, is a bit different. Even if the catalyst to engage as tenant possibly originates from the landscape beauty of Ōyamasenmaida—Hettinger[76] calls this 'aesthetic protectionism'—the initial stimulus for participation in the Ownership System changes later from a pure 'scenery obsession'[72] to a larger environmental view. Although for both, tenants and visitors, the landscape scenery ranges in the first rank, tenants consider the protection of the typical *satoyama* fauna significantly more important than the visitors do (30.29% versus 6.51%).

For the visitors, Ōyamasenmaida might 'only' be one destination among many (3/4 of the visitors know or have already visited similar landscapes before). The tenants, however, through their working commitment, their emotions, social contacts, the sounds and scents of the site, and their memories and thoughts became emotionally attached to this place. That is why they perceive Ōyamasenmaida as significantly more beautiful, more unique and a more interesting place. This finding was also observed by other scientists[77,78]: 'When we care about an object of appreciation, we become interested in it. Accordingly, this interest motivates us to look for the object's aesthetic value'[73]. The benefits and disadvantages of esthetically grounded environmentalism are discussed controversially in the scientific community[29,31,73,76,78,79], whereby it is often maintained that for sound environmentalism, nature appreciation needs to be serious and morally based and has to change from subjectivism to objectivism, from ego- or anthropocentrism to acentrism, from a scenery (and flagship-species) focus to an environmental focus[31,72]. However, we think that in order to attract people unconcerned with environmental issues, a less scientific but more emotional approach[80] would be a more promising way, also outside of Japan. 'Frequently, one of the main reasons for supporting environmental causes is to preserve beauty. Average citizens wish to preserve scenic landscapes, flowering plants, and attractive animals and will donate their money and sometimes their time to working toward such ends'[73]. This can also be shown in our study: 82.22% of the tenants would cease their participation, if the landscape of Ōyamasenmaida were a large, plain, mechanized paddy field. From this point of view, the bias for nature's beauty should not be dismissed as superficial, but rather seen as an opportunity to attract the interest of potential activists first. Later on, their esthetic appreciation could expand to ecological aspects. This has been successfully demonstrated in Ōyamasenmaida. Starting with a rural–urban exchange of people, collective cultivation and production of rice, soybeans, rice wine, arts and crafts lectures, and common feasting, the PA also tries to arouse the tenants' interest in ecological issues, e.g., the life cycle of fireflies.

Reflecting on the motivations driving voluntary activities for nature conservation, we can state that the tenants were motivated rather by needs belonging to Alderfer's categories[81] of growth (including the need for self-development and personal growth and advancement) and relatedness (interpersonal relationships and getting public recognition) than by the actual existence needs, such as quality food or food security.

Taming and cultivating is natural

The Japanese understanding of nature, a result of the interconnection between esthetics and religion since ancient times, has formed the conception of nature and man as a unity[34], where human influence is hardly viewed as an intrusion into nature. 'Only when nature is brought into the realm of the known, e.g. tamed, and there are some immediate personal gains, do most Japanese become interested in protecting nature'[32]. Accordingly, we assumed that the Japanese might consider a cultural landscape as untouched nature (hypothesis 3). We could verify this hypothesis, as more than one-quarter of the tenants and visitors perceive the terraced rice terraces of Ōyamasenmaida as untouched/very untouched nature. For this result, we also get support from the literature. Takeuchi reports that the traditional rural landscape in Japan (*satoyama*) is often regarded as original or untouched nature[4].

Sense of spirits in nature

'Measuring' religiosity or spirituality, particularly in a different cultural context, is difficult and can lead to wrong conclusions. As already mentioned in the Introduction, religiosity and/or spirituality have a different meaning in Japan as opposed to the Western world. In order to avoid this problem, our research question centered on the attitude of tenants and visitors toward nature spirits and we cautiously asked 'only' about the general acceptance of the existence of spirits in nature. It can be considered as a main result that 94.35% of the tenants and 90.22% of the visitors did not hesitate to answer this question—a sign that this query did not seem strange or non-sensical to them.

Most of what has been written by different authors up to now about Japanese religiosity reflects their own insights or imagination[46]. Empirical studies are still rather few[46]. Nevertheless, a Japanese scholar found out that out of 1800 respondents, 30.8% agreed '...that souls inhabit everything, such as mountains, rivers, grass, and trees'[46]. Our research question is very close to this. Also, in the Tenth Annual Japanese Character Study from 2000, cited by Stark et al., 59% of the interviewees answered that they feel that rivers and mountains have spirits[44]. In our study, 56.84% of the tenants and 50% of the visitors (always, often and sometimes) have the feeling that spirits exist in the mountains, valleys, rice terraces, orchards, streams, lakes, plants and trees. If we add the number of persons who have this feeling 'rarely', we even reach as much as 76.5% among the tenants and 72.89% among the visitors. This high rate of belief in nature spirits is surprising. However, the limitation of a cross-sectional study, namely the impossibility to distinguish between cause and effect due to a lack of observation over time, becomes particularly evident here, as we have no distinct explanation for the causality of the belief in spirits in nature among tenants and visitors. It could be possible that the special atmosphere (peace, silence, nature, nice landscape and, in the case of the tenants, handwork) of Ōyamasenmaida had an influence on the respondents' answers and that questioning of the same people, but in the loud and hectic conurbation of Tōkyō, would have led to different results (PC: Dr Birgit Staemmler, January 26, 2012).

Our hypothesis that tenants and visitors imagine differently the way in which nature is inhabited by spirits can be partly confirmed, as significant differences appear only in the category 'always'. Our assumption that, in contrast to the visitors, who just passed through, the feeling of nature spirits had developed more strongly among the tenants through their long-term involvement could not be confirmed in our research. With respect to age, tenants and visitors do not differ significantly, so that age difference as a reason for this phenomenon can also be excluded. One interpretation for our study could be that the tenants participate because they already have a certain tendency toward traditions and the will to return to nature, to the roots of Japanese culture (PC: Dr Birgit Staemmler, January 26, 2012). Cultivating rice is a good possibility to satisfy these feelings and the belief in nature spirits, in transcendence and otherworldliness is very close to this mental attitude (ibid.). The director of Ōyamasenmaida argues that the tenants became more receptive to *kami-sama* on account of producing their own food in the midst of 'great nature'.

Also unanswered is the reason for the highly significant difference between female tenants and female visitors in the category 'I always have the feeling of spirits in nature'. It can be assumed that among women a high spirituality appears more strongly in direct action (e.g., in participation in the Ownership System) than it is among men, so that the significant differences in our case become obvious only between the female tenants and female visitors.

Conclusions

The case of the Japanese Ownership System indicates that esthetic landscape appreciation can help mobilize an involvement in nature conservation activities. The interviewed volunteers in Ōyamasenmaida are rather more motivated by intrinsic reasons (most of all by the landscape esthetics, but also by nature experience, learning, hand work and socializing with like-minded) than by extrinsic ones (e.g., physical need to produce food, social pressure or approval from outside). Through able guidance (such as that provided by the PA in our case study) which also addresses emotion and fun—through traditional music, theatre and dance festivals, photo awards (for the yearly Ōyamasenmaida calendar), sport events (e.g., 'mud' volleyball competition in the rice fields), feasting (karaoke singing and harvest festivals), local food or handicraft workshops, renovation of old traditional houses for common purposes or collective nature experiences (fireflies, autumn moon and torch illuminated rice fields)—the volunteers could develop a broader ecological awareness. In consequence, their interest would also extend to other less attractive places. In Western countries (mainly in Central Europe or North America), nature conservation still works on a more scientific level, mobilizing volunteers through scientific evidence on, for example, losses of species or biodiversity. To also address the motivation of the volunteers on an emotional, esthetic or social level could be a promising way forward.

The extent to which spirituality is the cause for or an effect of active involvement in nature conservation activities cannot be derived from this survey. The subject of future studies could therefore be the question concerning the connection between spirituality (and religiosity) and the engagement in nature conservation activities. Even if a lot of international literature can be found related to the attitudes of religious persons in different fields of life—also those including environmental concerns—none of it addresses the question as to whether spiritual/religious motivation is the reason for engaging in nature conservation activities or whether the engagement itself results in spiritual/religious awareness.

Acknowledgements. The studies in Japan were financed by the German Academic Exchange Service (DAAD) and the Ministry of Education, Culture, Sports, Science, and Technology of Japan (MEXT). We wish to express our heartfelt gratitude to the Ōyamasenmaida Preservation Association, and in particular to Mitsuji Ishida, Yoshiko Sudo and Hitomi Taira, for their contribution and support during and after the fieldwork, as well as to all the tenants and visitors who completed the questionnaire survey. Special thanks are due to the persons who assisted in the translation process, namely to Kentaro Aoki (UNIDO/IIASA, Vienna), Nobuko Morishita (Institute of Oriental Manuscripts, Russian Academy of Sciences), Isabelle Prochaska (Department of Japanese Studies, University of Vienna), Nobuhiko Sawai (Gunma University, Graduate School of Medicine), Ayako Toko (WWF Japan) and Yuko Yoneda (Kyoto Prefectural University). We also wish to extend many thanks to Katharina Bardy of the Institute of Integrative Nature Conservation Research (BOKU-University) for her help with statistical analysis. We very much appreciated the advice of Birgit Staemmler (Department of Japanese Studies, University Tübingen) on Japanese religion and the kindness of Naoki Amako (Assistant Director, Biodiversity Policy Division, Nature Conservation Bureau, Ministry of the Environment, Japan) in allowing us to consult him on environmental data. In addition, we are indebted to Vivien Landauer for taking the time to edit the manuscript, and to the anonymous reviewers for providing their very helpful comments. Last, but not least, the corresponding author particularly would like to express her sincere thanks to Wolfgang Holzner (BOKU-University) for his kind suggestions, unwavering support and continual encouragement.

References

1 MAFF—Japanese Ministry of Agriculture, Forestry and Fisheries. 2010. FY2009 Annual Report on Food, Agriculture and Rural Areas in Japan. Summary. Available at Web site http://www.maff.go.jp/e/annual_report/2009/pdf/e_all.pdf (accessed February 9, 2012).

2 MOE—Japanese Ministry of the Environment, Nature Conservation Bureau (ed.). 2010. Biodiversity is Life. Biodiversity is our Life. The National Biodiversity Strategy of Japan 2010. Available at Web site http://www.biodic.go.jp/biodiversity/wakaru/library/files/nbsap2010/nbsap2010_EN.pdf (accessed August 14, 2012)

3 Takeuchi, K. 2001. Nature conservation strategies for the 'SATOYAMA' and 'SATOCHI', habitats for secondary nature in Japan. Global Environment Research 5(2):193–198.

4 Takeuchi, K. 2003. The Nature of Satoyama Landscapes. In K. Takeuchi, R.D. Brown, I. Washitani, A. Tsunekawa, and M. Yokohari (eds). Satoyama—The Traditional Rural Landscape of Japan. Springer-Verlag, Tokyo, p. 9–16.

5 FAO 2008. Conservation and Adaptive Management of Globally Important Agricultural Heritage Systems (GIAHS), Terminal Report. Project Symbol: UNTS/GLO/002/GEF, Project ID: 137561. FAO, Rome.

6 MAFF—Japanese Ministry of Agriculture, Forestry and Fisheries. 2009/2010. The 84th and 85th Statistical Yearbook of Ministry of Agriculture, Forestry and Fisheries (2008–2009 and 2009–2010). Available at Web site http://www.maff.go.jp/e/tokei/kikaku/nenji_e/nenji_index.html (accessed February 9, 2012).

7 MAFF—Ministry of Agriculture, Forestry and Fisheries, Japan. 2009. Chūsankanchiki nōgyō o meguru jōsei nōson shinkō kyoku (in Japanese). Available at Web site http://www.maff.go.jp/j/study/other/cyusan_taisaku/32/pdf/data1.pdf (accessed February 20, 2012).

8 Aizaki, H., Sato, K., and Osari, H. 2006. Contingent valuation approach in measuring the multifunctionality of agriculture and rural areas in Japan. Paddy Water Environment 4:217–222.

9 Iwabuchi, S., Kurechi, M., and Kashiwagi, M. 2010. Biodiversity in Rice Paddies 10th UN Convention on Biodiversity, Nagoya, 2010.

10 Eder, M. 1955. Die 'Reisseele' in Japan und Korea. Folklore Studies XIV:215–244 (in German).

11 Hagin Mayer, F. 1989. The calendar of village festivals: Japan. Asian Folklore Studies 48:141–147.

12 Kagawa-Fox, M. 2010. Environmental ethics from the Japanese perspective. Ethics, Place and the Environment, 13(1):57–73.

13 Oyadomari, M. 1989. Profiles. The rise and fall of the nature conservation movement in Japan in relation to some cultural values. Environmental Management 13(1):23–33.

14 Agency for Cultural Affairs, Japan, Department of Cultural Properties, Division of Monuments and Sites, Committee on

the Preservation, Development, and Utilization of Cultural Landscapes Associated with Agriculture, Forestry and Fisheries. 2003. The Report of the Study on the Protection of Cultural Landscapes Associated with Agriculture, Forestry and Fisheries. Available at Web site http://www.bunka.go.jp/english/pdf/nourinsuisan.pdf (accessed February 9, 2012).

15 MIC—Japanese Ministry of Internal Affairs and Communications of Japan, Statistics Bureau, Director-General for Policy Planning (Statistical Standards) and Statistical Research and Training Institute. 2010. Chapter 2. Population. In Statistical Handbook of Japan 2010. Available at Web site http://www.stat.go.jp/english/data/handbook/c02cont.htm#cha2_1 (accessed February 9, 2012).

16 MAFF—Ministry of Agriculture, Forestry and Fisheries, Japan. 2008. Annual Report on Food, Agriculture and Rural Areas in Japan FY 2008. Policies on Food, Agriculture and Rural Areas in Japan FY2007. Summary (Provisional Translation). Available at Web site http://www.maff.go.jp/e/annual_report/2008/pdf/e_all.pdf (accessed February 9, 2012).

17 Kobori, H. and Primack, R.B. 2004. Conservation for satoyama, the traditional landscape of Japan. Arnoldia 62(4):2–10.

18 Fukuda, H., Dyck, J., and Stout, J. 2003. Rice Sector Policies in Japan—Electronic Outlook Report from the Economic Research Service. Economic Research Service, USDA. p. 1–19. Available at Web site http://www.ers.usda.gov/publications/rcs/mar03/rcs030301/rcs0303-01.pdf (accessed December 1, 2011).

19 MAFF—Ministry of Agriculture, Forestry and Fisheries, Japan. 2009. Heisei 20 nendo. Chūsankanchiiki tō chokusetsushiharai seido no jisshijōkyō. Nōson shinkō kyoku (in Japanese). Available at Web site: http://www.maff.go.jp/j/nousin/tyusan/siharai_seido/pdf/h20_zissi_data3.pdf (accessed February 9, 2012).

20 Nakamichi, H. 2010. Rural revitalization through retirement farming in less-favored areas in Japan: The case of elderly farmers in Shikoku. In Conference Proceedings of the 4th International Conference of the Asian Rural Sociology Association (ARSA), Legazpi City, Philippines, September 6–10, 2010, p. 239–250.

21 MLIT—Japanese Ministry of Land, Infrastructure, Transport and Tourism. 2006 (in Japanese). Available at Web site http://www.mlit.go.jp/common/000029285.pdf (accessed February 9, 2012).

22 Nakagawa, S. 2003. Approaches to satoyama conservation. In K. Takeuchi, R.D. Brown, I. Washitani, A. Tsunekawa, and M. Yokohari (eds). Satoyama—The Traditional Rural Landscape of Japan. Springer-Verlag, Tokyo, p. 111–119.

23 Kuramoto, N. 2003. Citizen conservation of satoyama landscapes. In K. Takeuchi, R.D. Brown, I. Washitani, A. Tsunekawa, and M. Yokohari (eds). Satoyama—The Traditional Rural Landscape of Japan. Springer-Verlag, Tokyo, p. 23–39.

24 Takeuchi, K. 2003. National land planning of satoyama landscapes. In K. Takeuchi, R.D. Brown, I. Washitani, A. Tsunekawa, and M. Yokohari (eds). Satoyama—The Traditional Rural Landscape of Japan. Springer-Verlag, Tokyo, p. 200–208.

25 MAFF—Japanese Ministry of Agriculture, Forestry and Fisheries. 2010. The 85th Statistical Yearbook of Ministry of Agriculture, Forestry and Fisheries (2009–2010). Available at Web site http://www.maff.go.jp/e/tokei/kikaku/nenji_e/85nenji/index.html (accessed February 9, 2012).

26 Kobayashi, K. and Harada, C. 2010. Conservation of rice terraces in Japan – roles of the Sakaori rice terrace conservation association. Revija za geografijo – Journal for Geography 5–1:91–100.

27 Agency for Cultural Affairs, Department of Cultural Properties, Division of Monuments and Sites, Japan. 2003. Nihon no bunkatekikeikan. Nōrinsuisangyō ni kanren suru bunkatekikeikan no hogo ni kansuru chyōsakenkyū hōkokusho (in Japanese).

28 Kieninger, P.R., Yamaji, E., and Penker, M. 2011. Urban people as paddy farmers: the Japanese tanada ownership system discussed from a European perspective. Renewable Agriculture and Food Systems 26(4):328–341.

29 Saito, Y. 2002. Scenic national landscapes: Common themes in Japan and the United States. Essays in Philosophy 3(1): Article 5.

30 Kalland, A. and Asquith, P.J. 1997. Japanese perceptions of nature: Ideals and illusions. In P.J. Asquith and A. Kalland (eds). Japanese Images of Nature, Cultural Perspectives, Nordic Institute of Asian Studies, Man and Nature in Asia, No. 1. Curzon Press, Richmond, UK, p. 1–35.

31 Saito, Y. 2010. Future directions for environmental aesthetics. Environmental Values 19:373–391.

32 Kalland, A. 1995. Culture in Japanese nature. In O. Bruun and A. Kalland (eds). Asian Perceptions of Nature. A Critical Approach, Nordic Institute of Asian Studies, Studies in Asian Topics, No. 18. Curzon Press, Richmond, UK, p. 243–257.

33 Ashkenazi, M. 1997. The cannonization of nature in Japanese culture. Machinery of the natural in food modernization. In P.J. Asquith and A. Kalland (eds). Japanese Images of Nature, Cultural Perspectives, Nordic Institute of Asian Studies Man and Nature in Asia, No 1. Curzon Press, Richmond, UK, p. 206–220.

34 Nagasawa, K. 2008. The mountains, rivers, grasses and trees will all become Buddha – the Japanese view of nature and religion. In Biodiversity Network Japan (ed.). Conserving Nature. A Japanese Perspective, p. 32–35. Available at Web site: http://www.cbd.int/doc/external/cop-09/bnj-nature-en.pdf (accessed August 14, 2012).

35 Kieninger, P. and Holzner, W. 2011. Counting species or celebrating fireflies – concepts and conservation of nature in Europe and Japan. In S. Bergmann and H. Eaton (eds). Ecological Awareness – Exploring Religion, Ethics and Aesthetics. Studies in Religion and the Environment, 3. LIT Verlag Dr. W. Hopf, Berlin, p. 151–164.

36 Kellert, S.R. 1991. Japanese perception of wildlife. Conservation Biology 5:297–308.

37 Kellert, S.R. 1993. Attitudes, knowledge, and behavior toward wildlife among the industrial superpowers: United States, Japan, and Germany. Journal of Social Issues 49:53–69.

38 Kellert, S.R. 1995. Concepts of nature East and West. In M. Soule and G. Lease (eds). Reinventing Nature? Responses to Postmodern Deconstruction. Island Press, San Franciso, p. 103–121.

39 Asquith, P.J. and Kalland, A. 1997. Japanese Images of Nature, Cultural Perspectives. Nordic Institute of Asian Studies, Man and Nature in Asia, No 1. Curzon Press, Richmond, UK.
40 Kagami, M. 1998. Significance of the idea of walking in the woods as recreation in Japan. Gobal Environment Research 2:187–192.
41 Hayashi, A. 2002. Finding the voice of Japanese wilderness. International Journal of Wilderness 8:34–37.
42 Kohsaka, R. and Flintner, M. 2004. Exploring forest aesthetics using forestry photo contest: Case studies examining Japanese and German public preferences. Forest Policy and Economics 6:289–299.
43 Tanaka, K. 2010. Limitations for measuring religion in a different cultural context—the case of Japan. Social Science Journal 47:842–852.
44 Stark, R., Hamberg, E. and Miller, A.S. 2005. Exploring spirituality and unchurched religions in Amerika, Sweden, and Japan. Journal of Contemporary Religion 20(1):3–23.
45 Mullins, M.R. 2011. Religion in contemporary Japanese lives. In V. Lyon (ed.). Routledge Handbook of Japanese Culture and Society. Routledge, Oxford, p. 63–74.
46 Manabe, K. 2007. The structure of Japanese religiosity: Toward a re-examination of secularization in Japan. Kwansei Gakuin University Social Science Review 12:1–21.
47 Primack, R., Kobori, H., and Mori, S. 2000. Dragonfly pond restoration promotes conservation awareness in Japan. Conservation Biology 14:1553–1554.
48 Kobori, H. and Primack, R.B. 2003. Participatory conservation approaches for satoyama, the traditional forest and agricultural landscape of Japan. Ambio 32:307–311.
49 Horiuchi, M., Fukamachi, K., and Oku, H. 2011. Reed community restoration projects with citizen participation: An example of the practical use of satoyama landscape resources in Shiga Prefecture, Japan. Landscape Ecology Engineering 7:217–222.
50 Shibata, T. and Masuda, M. 2001. Sustainability of tanada owner system: A case study on Obasute Tanada, Koshoku City, Nagano Prefecture. Bulletin of Agricultural and Forestry Research Center, University of Tsukuba 14:19–28 (in Japanese, with English abstract).
51 Yamamoto, W., Yamaji, E., and Makiyama, M. 2001. Continuity of the *ownership* program of rice terraces from the viewpoint of participants' behavior—A case study of oyama-senmaida in Kamogawa City. Journal of Rural Planning Association 20:199–204.
52 Yamamoto, W., Yamaji, E., and Makiyama, M. 2002. Consciousness of rural people for ownership program of rice terraces. A case study of oyama-senmaida ownership program in Kamogawa City. Journal of Rural Planning Association 21:115–120 (in Japanese, with English abstract).
53 Yamamoto, W., Makiyama, M., and Yamaji, E. 2003. Continuity of the labor support of local farmers in 'ownership program' of rice terraces: Case study in Ohyama District, Kamogawa City. Journal of Rural Planning Association 22:112–121 (in Japanese, with English abstract).
54 Takao, K., Maeda, M., and Nonami, H. 2003. Residents' perception of procedural justice in implementing the rice terrace ownership system in Asuka Village, Nara Prefecture. Journal of Rural Planning Association 22:26–36 (in Japanese, with English abstract).
55 Yamaji, E. 2006. Enjoyment of rural amenities by ownership program of rice terraces. Journal of Rural Planning Association 25:206–212 (in Japanese).
56 Motonaka, M., Sasaki, K., and Aso, M. 2001. The cultural value and its conservation methodology of 'obasute, tagoto-no-tsuki' designated as a place of scenic beauty under the law for protection of cultural properties. Journal of the Japanese Institute of Landscape Architecture LRJ 64:475–478 (in Japanese, with English abstract).
57 Kurita, H., Kimura, Y., Matsumori, K., and Osari, H. 2004. A study on the relationship between physical features of terraced rice fields landscapes and their perception. Journal of Rural Planning Association 23:85–90 (in Japanese, with English abstract).
58 Aono, S., Kaga, H., Shimomura, Y., and Masuda, N. 2005. Study on landscape attractiveness to residents from the viewpoint of topographical features in agricultural area and the edge of senboku hill. Journal of the Japanese Institute of Landscape Architecture LRJ 68:753–756 (in Japanese, with English abstract).
59 Hiwasaki, L. 2005. Toward sustainable management of national parks in Japan: Securing local community and stakeholder participation. Environmental Management 35:753–764.
60 Natori, Y., Fukui, W., and Hikasa, M. 2005. Empowering nature conservation in Japanese rural areas: A planning strategy integrating visual and biological landscape perspectives. Landscape and Urban Planning 70:315–324.
61 Oku, H. and Fukamachi, K. 2006. The differences in scenic perception of forest visitors through their attributes and recreational activity. Landscape and Urban Planning 75:34–42.
62 Omotedani, A. and Murakami, S. 2006. A study of the spatial conditions for viewing the terraced paddy fields. Reports of the City Planning Institute of Japan 4:91–94.
63 Natori, Y. and Chenowet, R. 2008. Differences in rural landscape perceptions and preferences between farmers and naturalists. Journal of Environmental Psychology 28:250–267.
64 Iwata, Y., Fukamachi, K., and Morimoto, Y. 2010. Public perception of the cultural value of satoyama landscape types in Japan. Landscape Ecology Engineering 7(2):173–184.
65 Omura, H. 2004. Trees, forests and religion in Japan. Mountain Research and Development 24(2):179–182.
66 Yin, R.K. 2002. Case Study Research: Design and Methods. 3rd ed. Applied Social Research Methods Series, Volume 5. Sage Publications, Thousand Oaks, California.
67 Ōyamasenmaida Cultural Landscape Preservation Committee. 2006. Ōyama no Senmaida bunkatekikeikan hozon katsuyō keikaku. Kabushiki kaisha koa, Kamogawa Shi (in Japanese).
68 ŌSM—Ōyamasenmaida. 2012 (in Japanese). Available at Web site http://www.senmaida.com/index.php (accessed February 9, 2012).
69 Imperial Household Agency. 2011 (in Japanese). Available at Web site http://www.kunaicho.go.jp/activity/gonittei/01/h22gonittei-1-2010-3.html (accessed April 19, 2011).
70 Paier, D. 2010. Quantitative Sozialforschung. Eine Einführung, Facultas.wuv Universitätsverlag, Vienna (in German).

71. Merriam-Webster. 1992. The Merriam-Webster Dictionary of Synonyms and Antonyms. Merriam Webster, Springfield, MA.
72. Carlson, A. 2010. Contemporary environmental aesthetics and the requirements of environmentalism. Environmental Values 19:289–314.
73. Lintott, S. 2008. Towards Ecofriendly Aesthetics. In A. Carlson and S. Lintott (eds). Nature, Aesthetics, and Environmentalism. From Beauty to Duty. Columbia University Press, New York, p. 380–396.
74. Anderson, L.M. 1991. Land use designations affect perception of scenic beauty in forest landscapes. Forest Science 27(2):392–400.
75. Carlson, A. and Lintott, S. 2008. Introduction. Natural aesthetic value and environmentalism. In A. Carlson and S. Lintott (eds). Nature, Aesthetics, and Environmentalism. From Beauty to Duty. Columbia University Press, New York, p. 1–21.
76. Hettinger, N. 2008. Objectivity in environmental aesthetics and protection of the environment. In A. Carlson and S. Lintott (eds). Nature, Aesthetics, and Envi ronmentalism. From Beauty to Duty. Columbia University Press, New York, p. 413–437.
77. Ryan, R.L., Kaplan, R., and Grese, R. 2001. Predicting volunteer commitment in environmental stewardship programmes. Journal of Environmental Planning and Management 44(5):629–648.
78. Brady, E. 2008. Aesthetic character and aesthetic integrity in environmental conservation. In A. Carlson and S. Lintott (eds). Nature, Aesthetics, and Environmentalism. From Beauty to Duty. Columbia University Press, New York, p. 397–412.
79. Brady, E. 2006. Aesthetics in practice: Valuing the natural world. Environmental Values 15:227–291.
80. Kieninger, P., Holzner, W., and Kriechbaum, M. 2009. Biocultural diversity and satoyama. Emotions and the funfactor in nature conservation—a lesson from Japan. Bodenkultur 60(1):15–21.
81. Alderfer, C.P. 1969. An empirical test of a new theory of human needs. Organizational Behavior and Human Performance 4(2):142–175.

Curriculum vitae, publications and presentations

Rahmenschrift Dissertation Pia R. Kieninger

Curriculum vitae Dipl.-Ing. Dr.nat.techn.

Dipl.-Ing. Dr.nat.techn. Pia Regina Kieninger

University of Natural Resources and Life Sciences (BOKU)
Department of Integrative Biology and Biodiversity Research
Institute of Nature Conservation Research (INF)
Gregor-Mendelstr. 33, 1190 Vienna, Austria
Phone: +43-1-47654-4505/ Fax: +43-1-47654-4504
E-Mail: pia.kieninger@boku.ac.at
Homepage: http://www.dib.boku.ac.at/14858.html

Date of Birth: 30 August 1975. Nationality: German

Education

11/2011 – 02/2013	BOKU, Multimedia Services. Attended 'BOKUdoku' documentary seminar.
08/2006 – 10/2013	BOKU, Institute for Integrative Nature Conservation Research and Institute of Sustainable Economic Development/Regional Development Group. Doctoral thesis on 'Cultural landscape conservation through urban engagement in traditional land use management – A case study from Japan'.
04/2004 – 06/2006	University of Tokyo, Japan – Graduate School of Frontier Sciences, Institute of Environmental Studies, Laboratory of Biosphere Function. Visiting scientist.
10/2003 – 03/2004	BOKU, Centre for Environmental Studies and Nature Conservation. Start of doctoral course.
10/1999 – 07/2000	Polytechnic University of Milan, Italy – Department of Architecture and Urban Studies. Academic exchange.
09/1999	University for Foreigners Perugia, Italy. Intensive Italian language studies.

03/1996 – 02/2003	BOKU, studied Landscape Architecture and Planning. Graduated with distinction as Diplomingenieur (Dipl.-Ing.) (comparable to Master's degree).
11/1995 – 02/1996	University of Kassel, Germany – Landscape Architecture and Landscape Planning.

Scientific working positions

Since 03/2009	Scientific assistant at the BOKU, Institute for Integrative Nature Conservation Research.
10/2008 – 02/2009	Lecturer at the BOKU, Institute for Sustainable Economic Development.
10/2007 – 02/2009	Scientific project assistant at the BOKU, Institute for Nature Conservation Research and Institute for Botany.

Main fields of interest

- Civic engagement and group motivation for environmental conservation.
- Biocultural diversity and cultural landscape.
- Rural development.
- *Satoyama* (traditional rural landscape of Japan).
- Western and Eastern nature concepts.
- Sustainability and environmental protection.

Teaching experience (BOKU)

- Bachelor seminar/ Interdisciplinary project (in German) (since 2008).
- Excursions related to vegetation ecology (in German) (2009).
- Lecture on Science and Practice for Environment and Bio-resource Management (in German) (2009 – 2011).
- Lecture on Introduction to Agricultural Science (in German) (2009 – 2010).

- Lecture on the Basics of Agriculture as well as related excursions (in German) (2009 – 2011).
- Seminar on Biocultural Diversity in Rural Landscapes (in English) (since 2011).

Projects
- Phytodiversity and viticulture – Analysis and further development of agropolitical measures (current).
- River management plan Kleinsölkbach/Styria, Austria (current).
- Film project (documentary) on the avalanche disaster in the Nature Park Sölktäler 2010 (current).
- Goat-Grazing and biocultural diversity – A transdisciplinary system approach. How to optimize pastoral land use from an ecological and socioeconomic viewpoint. Joint project between the Institute for Nature Conservation Research, BOKU, and the Institute of Theology and Ecology at the Orthodox Academy of Crete (2010 – 2012).

Grants
- Grant from the BOKU Centre for Global Change and Sustainability to attend the 'BOKUdoku' documentary seminar (2011 – 2013).
- Grant from the Austrian Research Foundation (ÖFG) to attend the 'Conference of the Parties of the Convention on Biological Diversity' in Japan (October 2010)
- 'Monbukagakusho grant' of the Japanese Ministry of Education, Culture, Sports, Science, and Technology and the German Academic Exchange Service for scientific research in Japan (2004 – 2006).
- Grant from the Austrian Exchange Service and the European Community Action Scheme for the Mobility of University Students to support academic exchange (Erasmus) in Italy (1999 – 2000).

Commitment within the scientific community

- Member of the Society Sölktäler Nature Park (since 2013)
- Secretary of ForumL, Austrian Alumni Association of Landscape Planning and Architecture (2009–2013).
- Member of the Austrian Japan-Society for Science and Art (since 2012).
- Member of the BOKU Satoyama Platform (since 2010).
- Member of the Austrian Society of Agricultural Economics (since 2009).
- Member of the Austrian Agricultural Journalist's Association (since 2008).
- Member of the Ōyamasenmaida Preservation Association, Japan (since 2006).
- Reviewer for the Austrian Agency for International Cooperation in Education and Research (2013).
- Reviewer for the Yearbook of the Austrian Society of Agricultural Economy (2011).
- Member of the editorial board of zoll⊕, Austrian Magazine of Landscape and Open Space (since 2006).

Publications and presentations

I. Scientific articles

Kieninger, P.R., Penker, M., Yamaji, E. 2012. Esthetic and spiritual values motivating collective action for the conservation of traditional rural landscapes – A case study of rice terraces in Japan. Renewable Agriculture and Food Systems DOI: http://dx.doi.org/10.1017/S1742170512000269: 1–16.

Kieninger, P., Holzner, W. Andrianos, L. 2011. „Eine uralte Welt verschwindet vor unseren Augen." Pastorale Landnutzung einst und heute: Über die Almwirtschaft in den Sölktälern und den Weissen Bergen. zoll⊕ Österreichische Schriftenreihe für Landschaft und Freiraum 18: 80–86 (in German, with English abstract).

Kieninger, P.R., Yamaji, E., Penker, M. 2011. Urban people as paddy farmers: The Japanese Tanada Ownership System discussed from a European perspective. Renewable Agriculture and Food Systems 26 (4): 328–241.

Kieninger, P., Holzner, W., Kriechbaum, M. 2009. Biocultural Diversity and Satoyama. Emotions and the fun-factor in nature conservation – A lesson from Japan. Bodenkultur 60 (1): 15–21.

Kieninger, P., Penker, M. 2009. tanada-ownership-system. Kulturlandschaftserhaltung auf Japanisch. zoll⊕ Österreichische Schriftenreihe für Landschaft und Freiraum 14: 45–49 (in German, with English abstract).

Kieninger, P., Penker, M. 2009. Ehrenamtliches Engagement für die Kulturlandschaft. zoll⊕ Österreichische Schriftenreihe für Landschaft und Freiraum 14: 92–94 (in German, with English abstract).

II. Contributions to books

Kieninger, P., Holzner, W. 2011. Counting Species or Celebrating Fireflies – Concepts and Conservation of Nature in Europe and Japan. In: S. Bergmann, H. Eaton (eds). Ecological Awareness – Exploring Religion, Ethics and Aesthetics. Studies in Religion and the Environment, Vol. 3. LIT Verlag, Berlin–Münster, pp. 151–164.

Holzner, W., **Kieninger, P.**, Jahrl, I., Kriechbaum, M., Thaler, F. 2007. Die Alm auf dem Hochschneeberg als (Öko)- System. In: W. Holzner (ed.). Almen. Almwirtschaft und Biodiversität. Grüne Reihe des Lebensministeriums, Vol. 17. Böhlau Verlag, Vienna, pp. 219–263 (in German).

III. Popular science (a selection)

Kieninger, P., Holzner, W. 2012. Käfer kontra Förster – zum Umgang mit der Wildnis im Wald. zoll⊕ Österreichische Schriftenreihe für Landschaft und Freiraum 20: 23–29 (in German, with English abstract).

Kieninger, P., Penker, M., Holzner, W., Yamaji, E. 2010. Städter als Reisfeldpächter: Das Tanada Ownership System. Boku Insight 2: 13 (in German).

Kriechbaum, M., **Kieninger, P.**, Holzner, W. 2010. Steirerkas und Biodiversität. In: Dialog des Monats 05/10: Natur und Kultur – zwei Seiten einer Medaille? Guest commentary. Das Österreichische Nachhaltigkeitsportal, Lebensministerium. http://www.vielfaltleben.at/article/articleview/82577/1/26595 (verified 27 July 2012) (in German).

Kieninger, P. 2009. Auf den Wein warten...und warum Wartezeiten Sinn machen. zoll⊕ Österreichische Schriftenreihe für Landschaft und Freiraum 15: 36–40 (in German, with English abstract).

Kieninger, P. 2007. Ryokufū ni yuwarete – ōyamasenmaida no yume [Invited by the green wind – Ōyamasenmaida dreams]. In: Oyamasenmaida bulletin 34: 10–10 (in Japanese).

Kieninger, P. 2007. Japan: Kulturlandschaftspflege als Freizeitspaß – Andere Länder, gleiche Probleme. Ein Beispiel für Kulturlandschaftserhaltung in Fernost. Landschaften – Die Zeitung des Arbeitskreises Wachau 7: 6–6 (in German).

IV. Conference proceedings

Kieninger, P., Prochaska, I. 2010. Satoyama research at the University of Natural Resources and Applied Life Sciences (BOKU), Vienna. In: BIOLOG (ed.). Veranstaltungsdokumentation: Biodiversitätsforschung – Meilensteine zur Nachhaltigkeit. Biodiversity Research – Milestones for Sustainability. Wissenschaft und Praxis im Gespräch, 29.–30. März 2010, Berlin. Gießen, pp. 31–32.

Kieninger, P., Penker, M., Yamaji, E. 2009. Originelle Kulturlandschaftserhaltung im Nippon-Style: das tanada-ownership-system. In: I. Darnhofer, A. Grabner, J. Hambrusch, L. Kirner, A. Matscher, T. Oedl-Wieser, H. Peyerl, K.H. Pistrich, S. Pöchtrager, J. Reiter-Stelzl, M. Schermer, F. Sinabell (eds). Rollen der Landwirtschaft in benachteiligten Regionen, Tagungsband der 19. Jahrestagung der Österreichischen Gesellschaft für Agrarökonomie, Universität Innsbruck, 24.–25. September 2009. Facultas Verlag, Wien, pp. 31–32 (in German, with English abstract).

Kieninger, P., Holzner, W. 2009. The Impact of Ancient "Man/Nature" Concepts on Contemporary Nature Conservation Attitudes in Europe and Japan. In: L. Andrianos, K. Kenanidis, A. Papaderos (eds). Proceedings of the International Conference on Ecological Theology and Environmental Ethics – ECOTHEE, Orthodox Academy of Crete, 2–6 June, 2008. Kolympari, pp. 95–110.

Kieninger, P., Holzner, W., Kriechbaum, M. 2009. Emotions and the fun-factor in nature conservation – a lesson from Japan. In: B.E. Splechtna (ed.). Proceedings of the International Symposium on Preservation of Biocultural Diversity – a Global Issue, University of Natural Resources and Applied Life Sciences, Vienna – Center for Environmental Studies and Nature Conservation, May 6–8, 2008. Facultas Verlag, Wien, pp. 5–10.

V. Unpublished presentations and posters

Kieninger, P., Penker, M. 2011. Consumers as Voluntary Paddy Farmers – The Example of the Tanada Ownership System in Japan. XXIV European Congress for Rural Sociology – Inequality and Diversity in European Rural Areas, Mediterranean Agronomic Institute of Chania, Crete, Greece, 22–25 August 2011 (presentation).

Holzner, W., Kriechbaum, M., Splechtna, B., **Kieninger, P.** 2010. Klimaschutz und/oder Biodiversität in Kulturlandschaft? DIALOG Klimaschutzmaßnamen und Biodiversität. Impulsreferate und Podiumsdiskussion, Vienna, Austria, 15 September 2010 (presentation, in German).

Kieninger, P., Prochaska, I., Holzner, W. 2010. Edelweiß & Steirerkas – Biocultural Diversity in the Satoyama-Landscapes of the Austrian Alps. 10th Conference of the Parties (COP) of the Convention on Biological Diversity, Nagoya, Japan, 28–29 October 2010 (poster).

Kieninger, P., Holzner, W. 2009. SATOYAMA – the traditional rural landscape of Japan. Symposium: BOKU Naturschutz- und Biodiversitätsforschung, University of Natural Resource Management and Applied Life Sciences, Vienna, Austria, 16 October 2009 (poster).

Kieninger, P., Penker, M., Holzner, W., Yamaji, E. 2009. Städter als Reisfeldpächter: Das Tanada Ownership System. Symposium: BOKU Naturschutz- und Biodiversitätsforschung, University of Natural Resource Management and Applied Life Sciences, Vienna, Austria, 16 October 2009 (poster, in German).

i want morebooks!

Buy your books fast and straightforward online - at one of world's fastest growing online book stores! Environmentally sound due to Print-on-Demand technologies.

Buy your books online at
www.get-morebooks.com

Kaufen Sie Ihre Bücher schnell und unkompliziert online – auf einer der am schnellsten wachsenden Buchhandelsplattformen weltweit! Dank Print-On-Demand umwelt- und ressourcenschonend produziert.

Bücher schneller online kaufen
www.morebooks.de

VDM Verlagsservicegesellschaft mbH
Heinrich-Böcking-Str. 6-8
D - 66121 Saarbrücken

Telefon: +49 681 3720 174
Telefax: +49 681 3720 1749

info@vdm-vsg.de
www.vdm-vsg.de

Printed by Books on Demand GmbH, Norderstedt / Germany